AC/DC PRINCIPLES Workbook

AMERICAN TECHNICAL PUBLISHERS, INC.
HOMEWOOD, ILLINOIS 60430-4600

ATP Staff

1 2 3 4 5 6 7 8 9 – 07 – 9 8 7 6 5 4 3 2 1

Printed in the United States of America

ISBN 978-0-8269-1351-7

Contents

Introduction

AC/DC Principles Workbook is designed to reinforce information presented in *AC/DC Principles*. The textbook may be used as a reference to complete the learning activities in the workbook. Each chapter in the workbook covers information from the corresponding chapter in the textbook.

The question types used in the workbook include true-false, multiple choice, completion, identification, and calculations. For true-false questions, circle the T if the statement is true, or circle F if the statement is false. For multiple choice questions, write the letter of the correct answer in the answer blank next to the question. For completion questions, write the correct answer in the answer blank next to the question. For identification questions, write the letter of the correct corresponding answer in the answer blank next to the question. For calculation questions, write the correct answer in the space provided.

In addition to questions, activities are also included in each workbook chapter. The activities correlate with textbook chapter content and reinforce comprehension of related concepts and math principles.

Information presented in *AC/DC Principles Workbook* addresses common electrical theories and applications. Additional educational materials related to this and other topics is available from American Tech. To obtain information about all American Tech products, visit the American Tech web site at www.go2atp.com.

The Publisher

Basic Concepts of Electricity 1

Name ______________________________ Date ______________

True-False

T F **1.** Static electricity has some limited practical uses, such as in electrostatic spray-painting and electrostatic air filters.

T F **2.** Slower heat transfer occurs in materials that are better conductors of heat.

T F **3.** A conductor is a material that has a low electrical resistance and permits electrons to move through it easily.

T F **4.** The proton is the negatively charged particle in the nucleus of an atom.

T F **5.** The minimum permissible oxygen level is 12% to 14% by volume.

T F **6.** Multiplying the speed of an object by the distance that it moves equals the amount of work produced.

T F **7.** Because the gram is small and difficult to work with, the kilogram is sometimes used as a unit of measure when performing weight measurements.

T F **8.** To convert a Fahrenheit temperature reading to Celsius, 32 is added to the Fahrenheit reading, and the sum is divided by 1.8.

T F **9.** Voltage is also known as electromotive force.

T F **10.** The unit for energy and work is the watt.

T F **11.** The potential energy of a body equals the force times the distance it is raised.

T F **12.** Power is the rate of doing work or using energy.

T F **13.** In electrical systems, power loss is due to friction.

T F **14.** Ten to the fifth power (10^5) is equal to 1,000,000.

T F **15.** A gas is a matter that has a definite volume but no definite shape.

T F **16.** Copper carries more current for a given size than aluminum.

T F **17.** The hydrogen atom has the fewest protons of all atoms.

T F **18.** Pure silver is soft and subject to sulfidation when used for incorrect applications.

T F **19.** Frequent switching causes sulfidation of silver contacts.

T F **20.** Gold-plated contacts are used in high-current applications.

T F **21.** Germanium and silicon are used to make electronic components such as diodes and resistors.

T F **22.** The law of electric charges states that opposite charges repel and like charges attract.

T F **23.** Induction electrification occurs when a negatively charged object is brought close to a neutral object.

T F **24.** A positively charged atom is produced when there are a greater number of electrons than protons.

T F **25.** Some electrical equipment cannot be used with an aluminum conductor.

T F **26.** A grounding path must never be fused or switched and must be a permanent part of the electrical circuit.

T F **27.** Rubber insulating gloves or leather protectors can be used separately to achieve equal levels of protection from electrical shock hazards.

T F **28.** Holes and leaks in rubber insulating gloves are detected by filling each glove with water.

T F **29.** Class D fires are fires that include burning metals.

T F **30.** A space containing a process vessel is a confined space.

Multiple Choice

______________ **1.** The most widely used form of energy for transmission is ___.
A. electrical
B. chemical
C. mechanical
D. heat

________________ **2.** ___ is the most commonly used conductor.
A. Silver
B. Copper
C. Lead
D. none of the above

________________ **3.** ___ resistance is any force that tends to hinder the movement of an object.
A. Chemical
B. Electrical
C. Magnetic
D. Mechanical

________________ **4.** A factor that affects the resistance of a conductor is the ___ of the conductor.
A. length
B. conductor cross-sectional area
C. material
D. all of the above

________________ **5.** Conductors are commonly made from copper, aluminum, and ___.
A. steel
B. copper-clad aluminum
C. silver
D. zinc

________________ **6.** ___ is the formation of film on contact surfaces.
A. Agglomeration
B. Oxidation
C. Polymerization
D. Sulfidation

________________ **7.** Semiconductors are made from materials that have ___ valence electrons.
A. 2
B. 4
C. 8
D. 24

________________ **8.** Class ___ protective helmets protect against impact hazards and low-voltage shocks and burns.
A. B
B. C
C. E
D. G

__________________ **9.** Rubber insulating gloves with a green label are rated for a maximum use voltage of ___ V.
- A. 1000
- B. 17,000
- C. 26,500
- D. 36,000

__________________ **10.** Class ___ fires include burning electrical devices, motors, and transformers.
- A. A
- B. B
- C. C
- D. K

__________________ **11.** ___ is the process of charging an object.
- A. Induction
- B. Rectification
- C. Electrification
- D. Conduction

__________________ **12.** Derived units of measure have ___ relationships to base units.
- A. algebraic
- B. geometric
- C. metric
- D. none of the above

__________________ **13.** The ___ is used to define power.
- A. dyne
- B. erg
- C. joule
- D. newton

__________________ **14.** ___ power is the actual power used in an electrical circuit.
- A. Apparent
- B. Electrical
- C. Mechanical
- D. True

__________________ **15.** A(n) ___ is a material that has very high resistance to the flow of electrons.
- A. capacitor
- B. conductor
- C. insulator
- D. resistor

________________ **16.** The ___ is the negatively charged particle of an atom.
A. electron
B. neutron
C. proton
D. quark

________________ **17.** ___ electrification occurs any time two objects with different levels of charge make physical contact.
A. Automatic
B. Conduction
C. Contact
D. Induction

________________ **18.** As a standard, the ___ is used to protect people and property from hazards that arise from the use of electricity.
A. NEC®
B. NFPA
C. OSHA
D. UL

________________ **19.** The ___ is a national organization that provides guidance in assessing the hazards of the products of combustion.
A. NEC®
B. NFPA
C. OSHA
D. UL

________________ **20.** ___ is the process of placing a danger tag on a source of power.
A. Tagout
B. Lockout
C. PPE
D. none of the above

Completion

________________ **1.** A(n) ___ is the number of electrons passing a given point in one second.

________________ **2.** The base unit of time in the English and metric systems is the ___.

________________ **3.** ___ is the product of voltage and current in a circuit and is calculated without considering the phase shift that may be present between the voltage and current.

______________ **4.** The ___ of a system is the ratio of the power output to the power input.

______________ **5.** A(n) ___ is a material that exhibits an electrical conductivity between that of a conductor (high conductivity) and that of an insulator (low conductivity).

______________ **6.** A(n) ___ is a substance that cannot be chemically broken down and contains atoms of only one variety.

______________ **7.** There are ___ elements, 92 of which are natural.

______________ **8.** ___ is a condition that results any time a body becomes part of an electrical circuit.

______________ **9.** Class ___ protective helmets protect against high-voltage shock and burns, impact hazards, and penetration by falling or flying objects.

______________ **10.** Fire extinguishers are selected for the class of fire based on the ___ of the material.

______________ **11.** ___ is the measurement of matter contained in an object.

______________ **12.** ___ current flow is typically used when designing solid-state circuits and systems.

______________ **13.** The Greek symbol "Ω" represents the ___.

______________ **14.** A(n) ___ is equal to one joule per second in the metric system.

______________ **15.** ___ contacts are used in high-current applications because of their high melting temperature.

______________ **16.** ___ electrification occurs when a conductor connects two wires with different charges.

______________ **17.** ___ safety goggles protect against low-voltage arc hazards.

______________ **18.** Rubber insulating matting is used to protect electricians when working with voltages above ___ V.

______________ **19.** A(n) ___ has the same importance as a lock and is used alone only when a lock does not fit the disconnect device.

______________ **20.** Fuel, heat, and ___ are required to start and sustain a fire.

Identification—Safety Equipment

__________________ **1.** Rubber insulating gloves

__________________ **2.** Insulating matting

__________________ **3.** Protective helmet

__________________ **4.** Safety glasses

__________________ **5.** Earplugs

__________________ **6.** Safety shoes

__________________ **7.** Fire-resistant clothing

__________________ **8.** Leather gloves

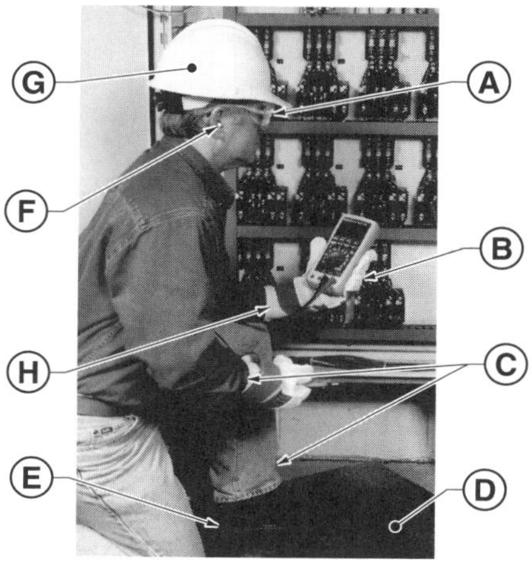

G A F B H C E D

Elements

Using the Chemical Elements table on the last page of the chapter, write the chemical symbol for each element.

__________________ **1.** Oxygen = ___

__________________ **2.** Copper = ___

__________________ **3.** Hydrogen = ___

__________________ **4.** Iron = ___

__________________ **5.** Aluminum = ___

__________________ **6.** Silver = ___

__________________ **7.** Gold = ___

__________________ **8.** Helium = ___

__________________ **9.** Sodium = ___

__________________ **10.** Argon = ___

__________________ **11.** Neon = ___

__________________ **12.** Tungsten = ___

__________________ **13.** Cadmium = ___

__________________ **14.** Chlorine = ___

________________ **15.** Carbon = ___

________________ **16.** Silicon = ___

________________ **17.** Germanium = ___

________________ **18.** Nickel = ___

________________ **19.** Mercury = ___

________________ **20.** Lead = ___

Fire Classes

Determine the class of fire for each of the following figures.

________________ **1.** Fire 1 is a class ___ fire.

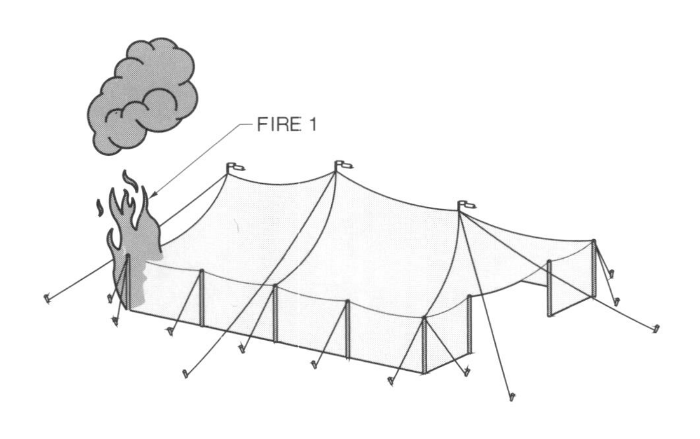

________________ **2.** Fire 2 is a class ___ fire.

_______________ **3.** Fire 3 is a class ___ fire.

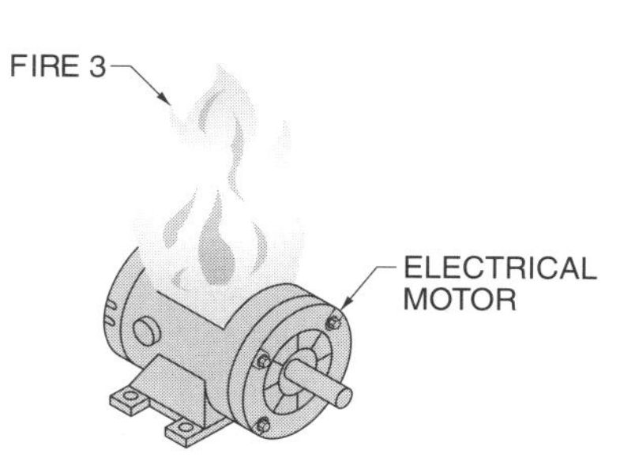

_______________ **4.** Fire 4 is a class ___ fire.

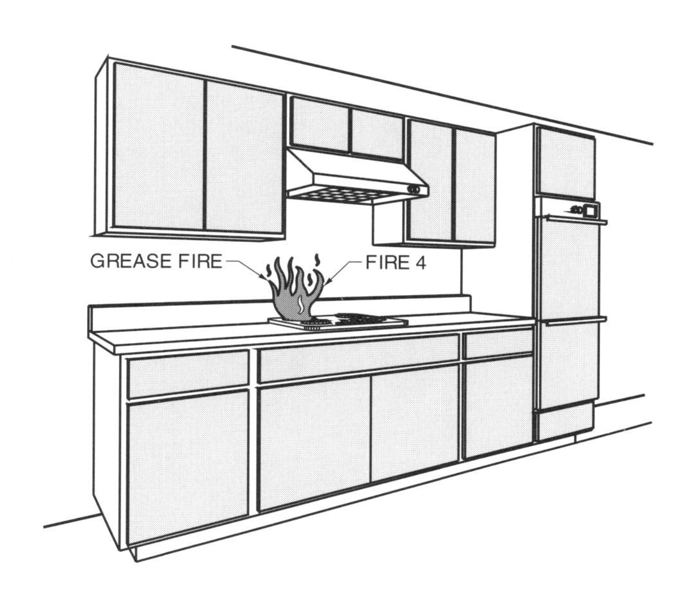

CHEMICAL ELEMENTS

Name	Symbol	Valence Electrons	Atomic Weight*	Atomic Number	Name	Symbol	Valence Electrons	Atomic Weight*	Atomic Number
Actinium	Ac	2	[227]	89	Neon	Ne	8	20.183	10
Aluminum	Al	3	26.9815	13	Neptunium	Np	2	[237]	93
Americium	Am	2	[243]	95	Nickel	Ni	2	58.71	28
Antimony	Sb	5	121.75	51	Niobium	Nb	1	92.906	41
Argon	Ar	8	39.948	18	Nitrogen	N	5	14.0067	7
Arsenic	As	5	74.9216	33	Nobelium	No	2	[255]	102
Astatine	At	7	[210]	85	Osmium	Os	2	190.2	76
Barium	Ba	2	137.34	56	Oxygen	O	6	15.9994	8
Berkelium	Bk	2	[247]	97	Palladium	Pd	—	106.4	46
Beryllium	Be	2	9.0122	4	Phosphorus	P	5	30.9738	15
Bismuth	Bi	5	208.980	83	Platinum	Pt	1	195.09	78
Boron	B	3	10.811	5	Plutonium	Pu	2	[244]	94
Bromine	Br	7	79.909	35	Polonium	Po	6	[210]	84
Cadmium	Cd	2	112.40	48	Potassium	K	1	39.102	19
Calcium	Ca	2	40.08	20	Praseodymium	Pr	2	140.907	59
Californium	Cf	2	[251]	98	Promethium	Pm	2	[145]	61
Carbon	C	4	12.01115	6	Protactinium	Pa	2	[231]	91
Cerium	Ce	2	140.12	58	Radium	Ra	2	[226]	88
Cesium	Cs	1	132.905	55	Radon	Rn	8	[222]	86
Chlorine	Cl	7	35.453	17	Rhenium	Re	2	186.2	75
Chromium	Cr	1	51.996	24	Rhodium	Rh	1	102.905	45
Cobalt	Co	2	58.9332	27	Rubidium	Rb	1	85.47	37
Copper	Cu	1	63.54	29	Ruthenium	Ru	1	101.07	44
Curium	Cm	2	[247]	96	Samarium	Sm	2	150.35	62
Dysprosium	Dy	2	162.50	66	Scandium	Sc	2	44.956	21
Einsteinium	Es	2	[254]	99	Selenium	Se	6	78.96	34
Erbium	Er	2	167.26	68	Silicon	Si	4	28.086	14
Europium	Eu	2	151.96	63	Silver	Ag	1	107.870	47
Fermium	Fm	2	[257]	100	Sodium	Na	1	22.9898	11
Fluorine	F	7	18.9984	9	Strontium	Sr	2	87.62	38
Francium	Fr	1	[223]	87	Sulfur	S	6	32.064	16
Gadolinium	Gd	2	157.25	64	Tantalum	Ta	2	180.948	73
Gallium	Ga	3	69.72	31	Technetium	Tc	2	[97]	43
Germanium	Ge	4	72.59	32	Tellurium	Te	6	127.60	52
Gold	Au	1	196.967	79	Terbium	Tb	2	158.924	65
Hafnium	Hf	2	178.49	72	Thallium	Tl	3	204.37	81
Helium	He	2	4.0026	2	Thorium	Th	2	232.038	90
Holmium	Ho	2	164.930	67	Thulium	Tm	2	168.934	69
Hydrogen	H	1	1.00797	1	Tin	Sn	4	118.69	50
Indium	In	3	114.82	49	Titanium	Ti	2	47.90	22
Iodine	I	7	126.9044	53	Tungsten	W	2	183.85	74
Iridium	Ir	2	192.2	77	Unnilennium	Une	2	[266]	109
Iron	Fe	2	55.847	26	Unnilhexium	Unh	2	[263]	106
Krypton	Kr	8	83.80	36	Unniloctium	Uno	—	[265]	108
Lanthanum	La	2	138.91	57	Unnilpentium	Unp	2	[262]	105
Lawrencium	Lr	2	[256]	103	Unnilquadium	Unq	2	[261]	104
Lead	Pb	4	207.19	82	Unnilseptium	Uns	2	[262]	107
Lithium	Li	1	6.939	3	Uranium	U	2	238.03	92
Lutetium	Lu	2	174.97	71	Vanadium	V	2	50.942	23
Magnesium	Mg	2	24.312	12	Xenon	Xe	8	131.30	54
Manganese	Mn	2	54.9380	25	Ytterbium	Yb	2	173.04	70
Mendelevium	Md	2	[258]	101	Yttrium	Y	2	88.905	39
Mercury	Hg	2	200.59	80	Zinc	Zn	2	65.37	30
Molybdenum	Mo	1	95.94	42	Zirconium	Zr	2	91.22	40
Neodymium	Nd	2	144.24	60					

* a number in brackets indicates the mass number of the most stable isotope

Resistance 2

Name ______________________________ Date ______________

True-False

T F 1. Ohm's law is the relationship between voltage, power, and resistance in an electrical circuit.

T F 2. The resistance of a conductor is directly proportional to its length.

T F 3. The smaller the AWG value, the smaller the cross-sectional area and the lighter the wire.

T F 4. Cermet is a mixture of fine particles of glass or ceramic and powdered metals such as silver, platinum, or gold.

T F 5. Power resistors are not available in adjustable designs.

T F 6. The conductor with the lowest resistivity is silver.

T F 7. The electrical unit of resistance is the joule.

T F 8. Resistivity can be calculated using either the metric or English system.

T F 9. Tin is a better conductor than nickel.

T F 10. Most metals have about a 4% increase in resistance for every 10°C increase in temperature.

T F 11. Silver has a higher temperature coefficient than aluminum.

T F 12. A fixed resistor can have two different resistance values depending on the application.

T F 13. Standard composition of stranded conductors can be 7, 19, or 37 strands.

T F 14. Resistors are normally described by electrical resistance, tolerance, power rating, method of construction, and circuit application.

T F 15. No relationship exists between a resistor's resistance value and its power rating.

T F **16.** The resistance value of carbon composition resistors changes very little with age.

T F **17.** On a four-color band resistor, the fourth band represents the multiplier.

T F **18.** The resistance value of metal film resistors changes very little with age.

T F **19.** Potentiometers are normally used as voltage dividers.

T F **20.** Resistors can be used to reduce the source voltage to the required voltage for a given load.

T F **21.** Continuity checkers can be used on both low- and high-resistance circuits.

T F **22.** An analog multimeter relies on circuit power to obtain a measurement.

T F **23.** A megohmmeter is used to detect insulation deterioration by measuring high resistance under high-voltage conditions.

Multiple Choice

______________ **1.** Resistivity is represented in mathematical equations by the Greek letter ___.

A. alpha (α)
B. lambda (λ)
C. omega (Ω)
D. rho (ρ)

______________ **2.** Relative resistance is the comparison of the resistance of a given material to the resistance of ___.

A. aluminum
B. copper
C. gold
D. silver

______________ **3.** Temperature coefficient is represented in mathematical equations by the Greek letter ___.

A. alpha (α)
B. lambda (λ)
C. rho (ρ)
D. sigma (σ)

____________ **4.** A 10 AWG wire has a resistance of ___ Ω per 1000′.
A. 1
B. 1.5
C. 5
D. 10

____________ **5.** The least expensive type of resistor is the ___ resistor.
A. wire wound
B. metal film
C. chip resistors
D. carbon composition

____________ **6.** A resistor with color bands (as read from left to right) of blue, green, brown, and gold has a value of ___ Ω at a tolerance of ±5%.
A. 6.5
B. 65
C. 650
D. 6500

____________ **7.** When applicable, the fifth color band on a resistor represents ___.
A. quality
B. temperature coefficient
C. military specification
D. polarity

____________ **8.** Potentiometers have ___ terminals, while rheostats have ___ terminals.
A. 2; 3
B. 2; 4
C. 3; 2
D. 4; 2

____________ **9.** A DMM normally measures resistance as low as ___ Ω and high as ___ MΩ.
A. 0.01; 300
B. 0.1; 300
C. 0.01; 500
D. 0.1; 500

____________ **10.** The resistivity of copper is ___ Ω/cmil/ft.
A. 5.1
B. 9.6
C. 10.4
D. 17.1

__________ **11.** The four factors affecting the resistance of a substance are type of material, conductor length, conductor cross-sectional area, and ___.
- A. type of application
- B. temperature
- C. color
- D. altitude

__________ **12.** Good insulation materials, such as glass and mica, have relative resistance at least ___ times that of copper.
- A. 1000
- B. 10,000
- C. 100,000
- D. 1,000,000

__________ **13.** In the SI system, conductance is measured in ___.
- A. ampere/turns (A/t)
- B. kilovolts (kV)
- C. ohms (Ω)
- D. siemens (S)

__________ **14.** Relative conductance is the ability of a specific conductor to carry electrons as compared to the ability of a(n) ___ conductor to carry electrons.
- A. aluminum
- B. copper
- C. gold
- D. silver

__________ **15.** The ___ is the unit of energy measurement.
- A. ampere (A)
- B. ohm (Ω)
- C. joule (J)
- D. volt (V)

__________ **16.** Resistors with no tolerance band have a tolerance of ±___%.
- A. 0.01
- B. 2
- C. 5
- D. 20

__________ **17.** If a range of voltages is required, a ___ can be used in place of a resistor to provide an adjustable output voltage.
- A. ballast resistor
- B. capacitor
- C. potentiometer
- D. rheostat

________ **18.** Megohmmeters are often referred to as ___ or insulation resistance testers.
A. Meggers®
B. DMMs
C. continuity checkers
D. voltmeters

________ **19.** A(n) ___ is an electromechanical device that indicates readings by the mechanical motion of a pointer.
A. analog multimeter
B. DMM
C. megohmmeter
D. continuity checker

Completion

________ **1.** ___ is the resistance of a material of a specific cubic size.

________ **2.** ___ is the ability of voltage to produce electron flow through a resistance.

________ **3.** Resistance of a conductor with a negative temperature coefficient ___ with an increase in temperature.

________ **4.** A(n) ___ is a material that exhibits zero resistance and imposes no opposition to electron flow nor consumes power at temperatures close to absolute zero (0° K, –273.15°C, –459.67°F).

________ **5.** ___ is the maximum amount of current a conductor can carry continuously without exceeding its temperature rating.

________ **6.** A carbon composition resistor is a resistor that is constructed using carbon graphite mixed with ___.

________ **7.** The tolerance rating of resistors used in electrical measurement instruments is usually ±___% or less.

________ **8.** Metal film resistors have replaced ___ resistors for applications that require a high degree of accuracy.

________ **9.** ___ resistors are used in high-wattage circuits where the need to make an adjustment is infrequent.

________ **10.** ___ resistors are used to compensate for changes of voltages in power lines that are connected to electronic equipment sensitive to fluctuations.

________ **11.** ___ resistors are normally manufactured with a hollow center so that air flowing through them can help dissipate heat.

_______________ **12.** An arrow on any electrical symbol indicates the ___ of the device.

_______________ **13.** A(n) ___ consists of a block of metal that transfers the heat from the resistor to the chassis.

_______________ **14.** ___ are resistors that act as voltage dividers.

_______________ **15.** A(n) ___ resistor is a resistor used to discharge a capacitor.

_______________ **16.** A(n) ___ circuit is a circuit that has only one current path.

_______________ **17.** A(n) ___ circuit is a circuit that has more than one current path.

_______________ **18.** A(n) ___ is an instrument that indicates an open or closed circuit in a circuit in which all power is OFF.

_______________ **19.** A(n) ___ error is the inaccuracy created by the difference in apparent position of an analog meter pointer when viewed from different angles not directly perpendicular to the pointer and scale.

_______________ **20.** A(n) ___ DMM is a DMM that automatically adjusts to a higher range if its initial range is not high enough.

_______________ **21.** A(n) ___ temperature coefficient is an increase in the resistance of a material with an increase in temperature.

Identification—Resistor Types

_______________ **1.** Carbon composition resistor

_______________ **2.** Metal film resistor

_______________ **3.** Cermet film resistor

_______________ **4.** Fixed power resistor

_______________ **5.** Adjustable power resistor

_______________ **6.** Tapped power resistor

_______________ **7.** Resistor networks

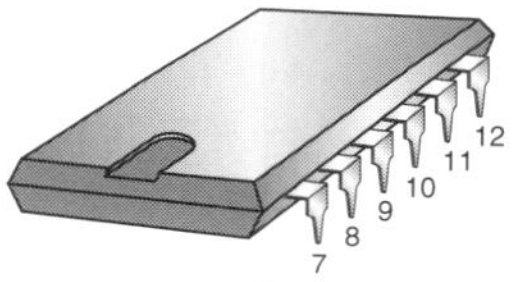

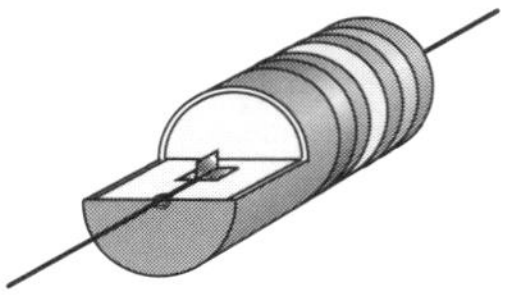

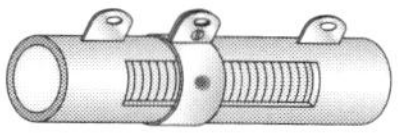

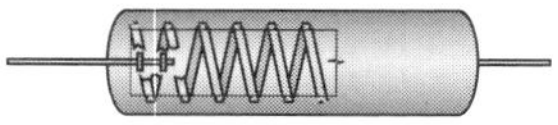

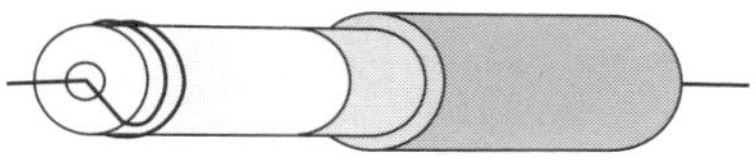

Identification—Digital Multimeter Functions

______________ **1.** AC current

______________ **2.** AC voltage

______________ **3.** AC and DC current

______________ **4.** AC and DC voltage

______________ **5.** Access blue functions

______________ **6.** Access yellow functions

______________ **7.** Capacitance

______________ **8.** Digital display

______________ **9.** Function switch

______________ **10.** Resistance/continuity check setting

______________ **11.** Temperature

______________ **12.** Test lead connection jacks

Calculating Total Resistance of Resistors Connected in Series

Calculate the total resistance of each series circuit.

______________ **1.** R_T = ___ Ω

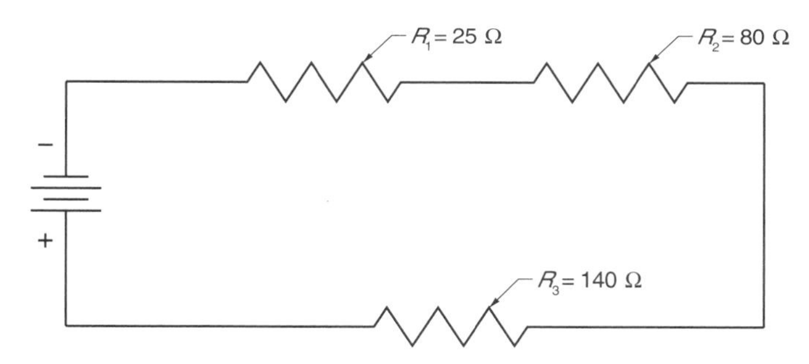

______________ **2.** R_T = ___ kΩ

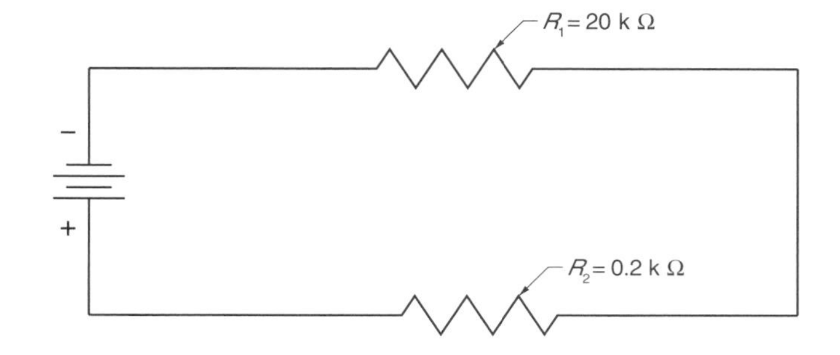

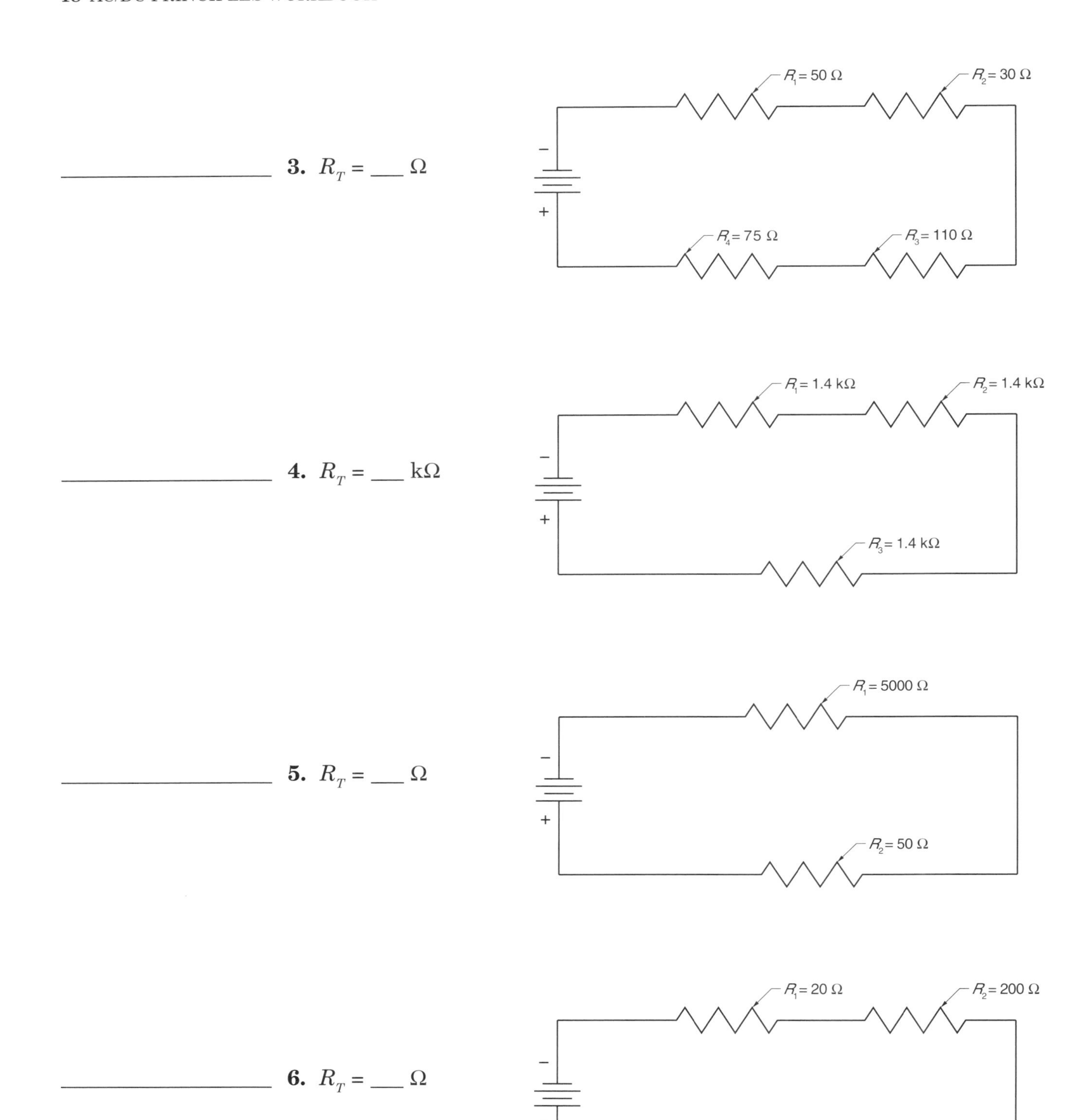

______________ **3.** R_T = ___ Ω

______________ **4.** R_T = ___ kΩ

______________ **5.** R_T = ___ Ω

______________ **6.** R_T = ___ Ω

Calculating Total Resistance of Resistors Connected in Parallel

Calculate the total resistance of each parallel circuit.

______________ **1.** R_T = ___ Ω

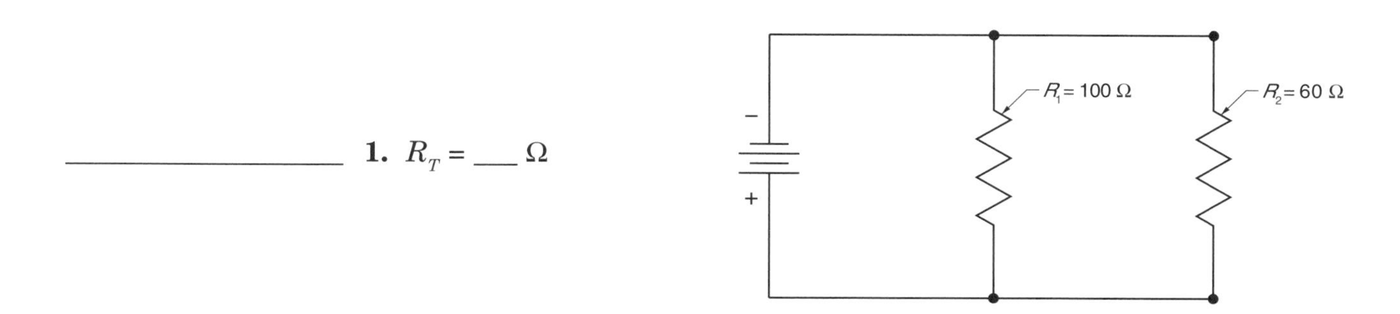

______________ **2.** R_T = ___ Ω

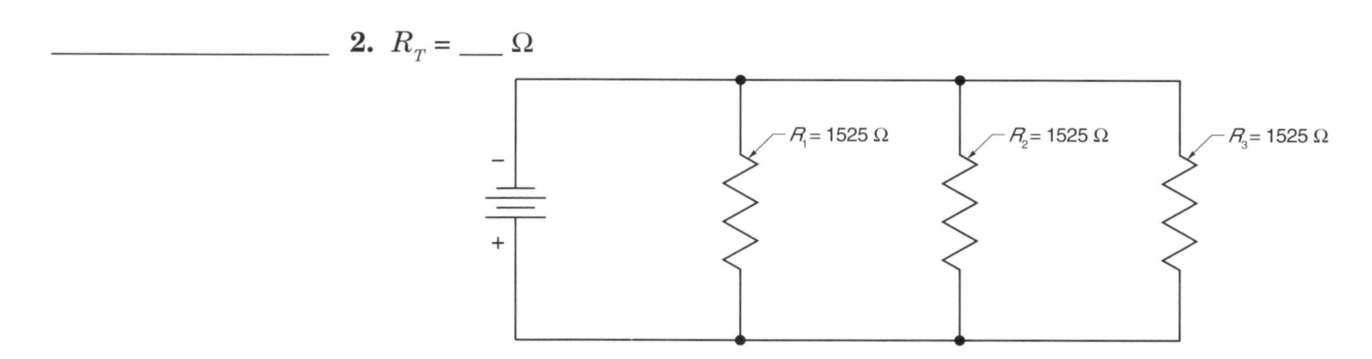

______________ **3.** R_T = ___ Ω

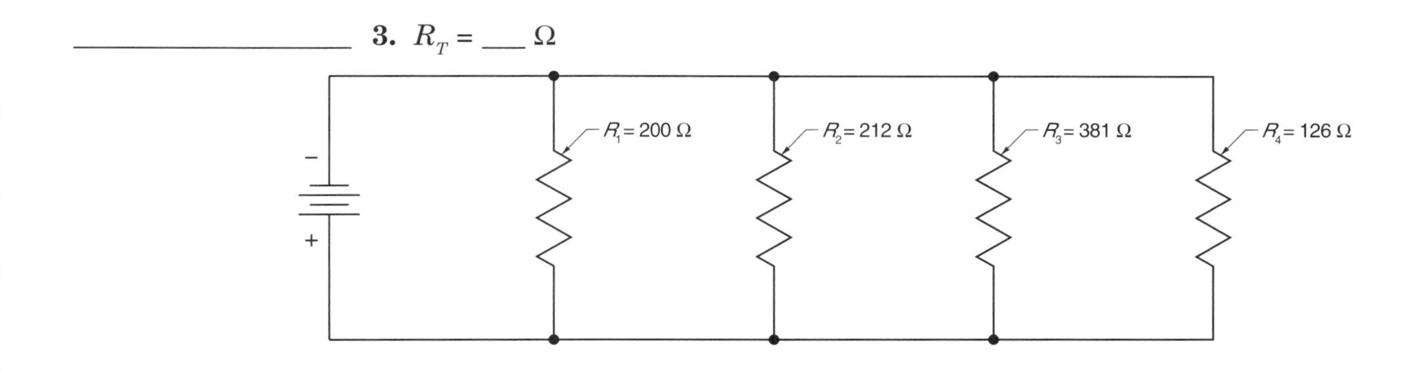

______________ **4.** R_T = ___ Ω

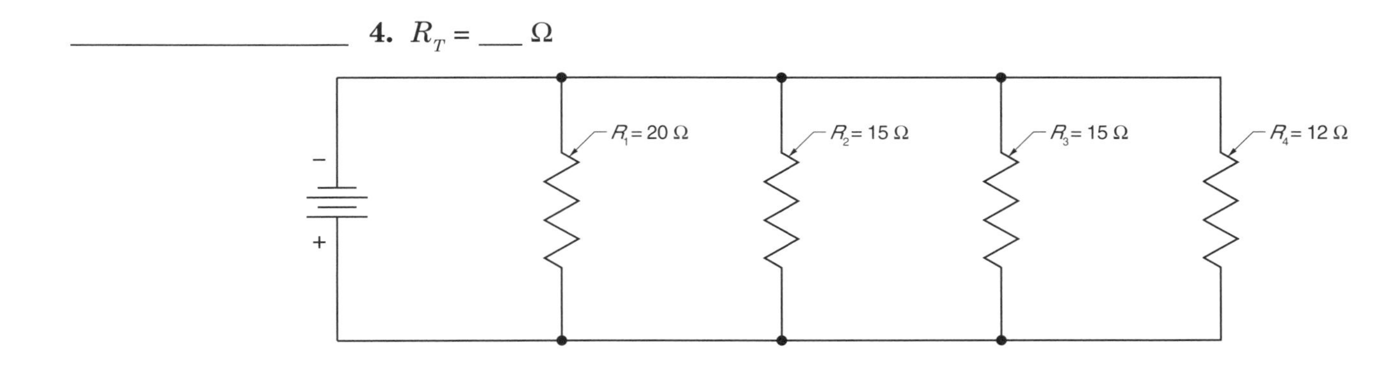

______________ **5.** R_T = ___ Ω

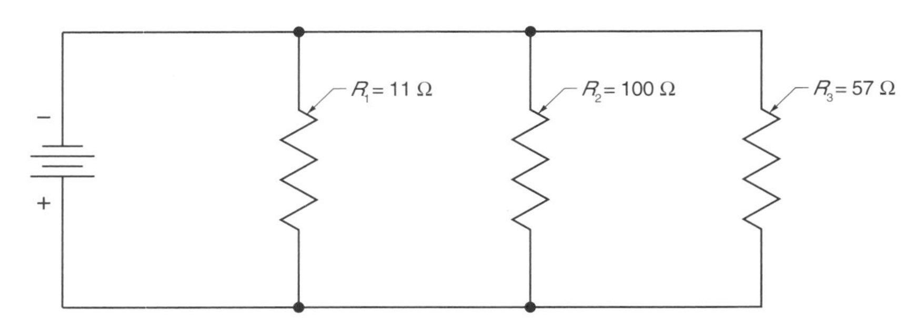

______________ **6.** R_T = ___ Ω

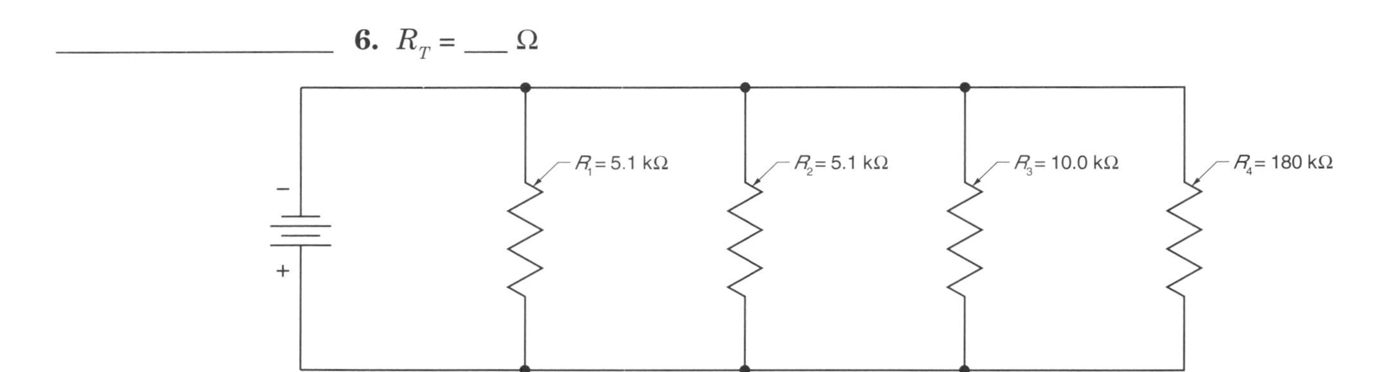

Resistor Color Coding

Give the resistance, tolerance, and resistance in simplest form using the Resistor Color Codes chart.

RESISTOR COLOR CODES				
Color	Digit		Multiplier	Tolerance*
	1st	2nd		
Black (BK)	0	0	1	0
Brown (BR)	1	1	10	—
Red (R)	2	2	100	—
Orange (O)	3	3	1000	—
Yellow (Y)	4	4	10,000	—
Green (G)	5	5	100,000	—
Blue (BL)	6	6	1,000,000	—
Violet (V)	7	7	10,000,000	—
Gray (GY)	8	8	100,000,000	—
White (W)	9	9	1,000,000,000	—
Gold (Au)	—	—	0.1	5
Silver (Ag)	—	—	0.01	10
None	—	—	0	20
	BAND 1	BAND 2	BAND 3	BAND 4

* in %

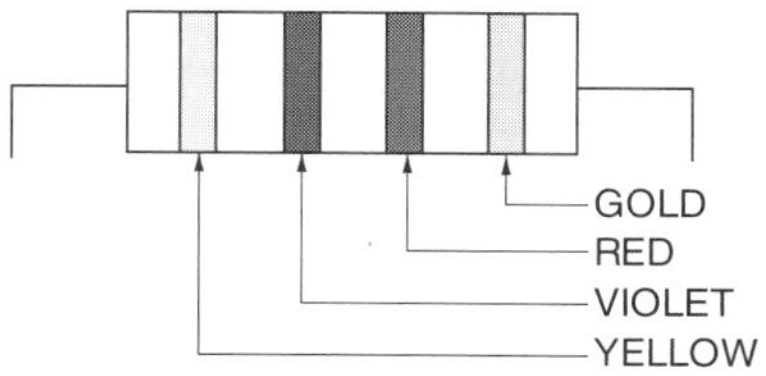

________________ **1.** Resistance is ___ Ω.

________________ **2.** Tolerance is ±___%.

________________ **3.** The simplest form is ___.

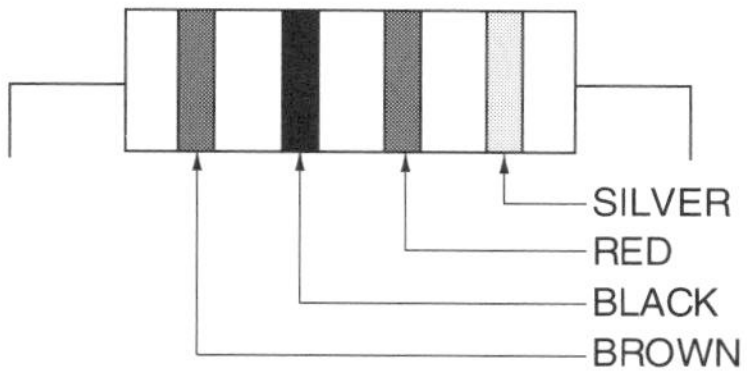

GOLD
RED
VIOLET
YELLOW

________________ **4.** Resistance is ___ Ω.

________________ **5.** Tolerance is ±___%.

________________ **6.** The simplest form is ___.

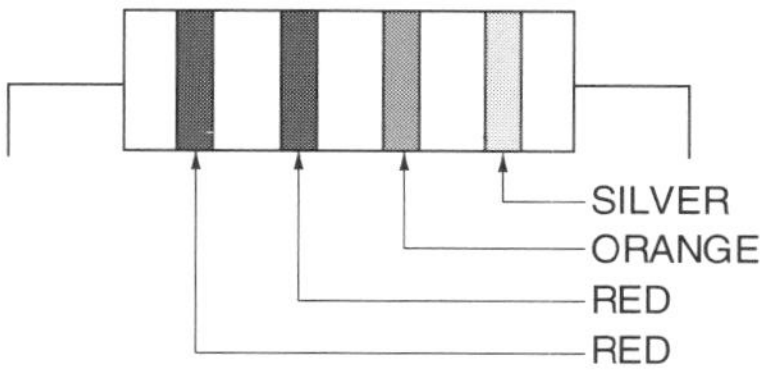

SILVER
RED
BLACK
BROWN

________________ **7.** Resistance is ___ Ω.

________________ **8.** Tolerance is ±___%.

________________ **9.** The simplest form is ___.

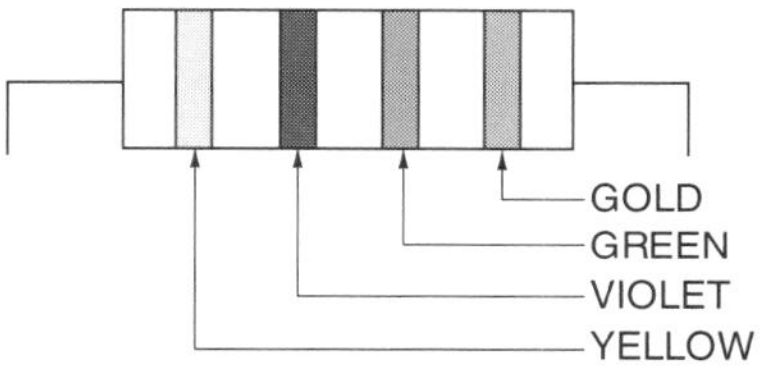

SILVER
ORANGE
RED
RED

________________ **10.** Resistance is ___ Ω.

________________ **11.** Tolerance is ±___%.

________________ **12.** The simplest form is ___.

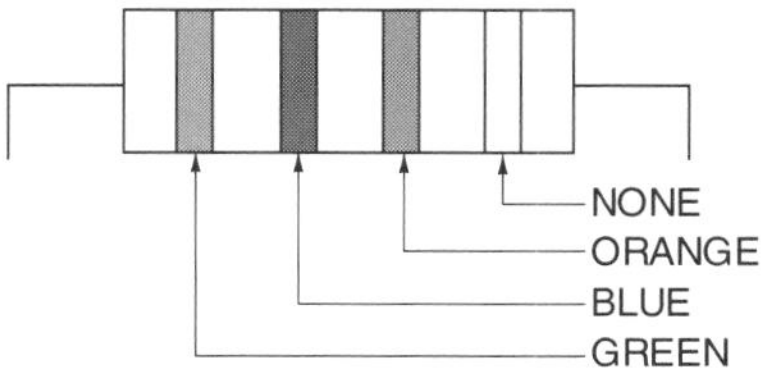

GOLD
GREEN
VIOLET
YELLOW

________________ **13.** Resistance is ___ Ω.

________________ **14.** Tolerance is ±___%.

________________ **15.** The simplest form is ___.

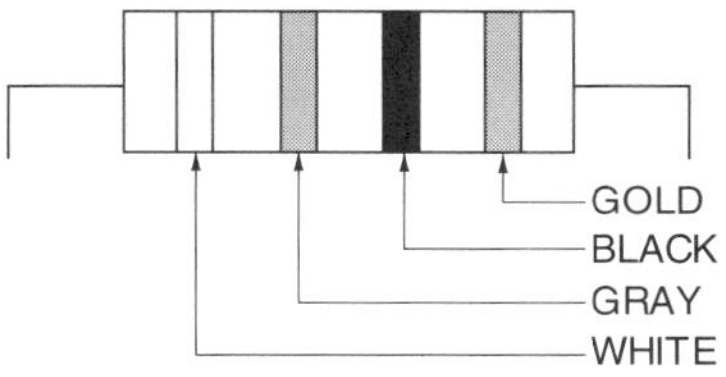

NONE
ORANGE
BLUE
GREEN

________________ **16.** Resistance is ___ Ω.

________________ **17.** Tolerance is ±___%.

________________ **18.** The simplest form is ___.

GOLD
BLACK
GRAY
WHITE

__________________ **19.** Resistance is ___ Ω.

__________________ **20.** Tolerance is ±___%.

__________________ **21.** The simplest form is ___.

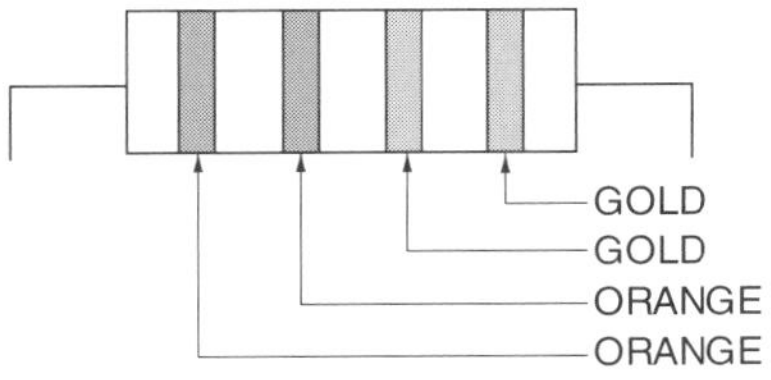

GOLD
GOLD
ORANGE
ORANGE

__________________ **22.** Resistance is ___ Ω.

__________________ **23.** Tolerance is ±___%.

__________________ **24.** The simplest form is ___.

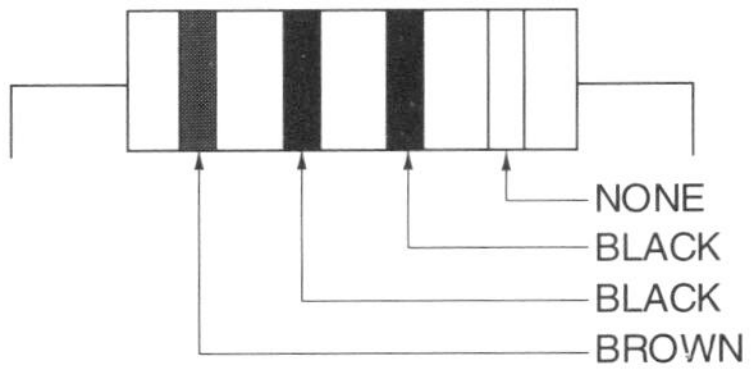

NONE
BLACK
BLACK
BROWN

Voltage Sources 3

Name ______________________ Date ____________

True-False

T F 1. Lightning rods are designed to take a direct hit from lightning strokes.

T F 2. A photovoltaic cell is sometimes known as a solar cell.

T F 3. Sodium and potassium release electrons when exposed to light.

T F 4. A battery is an AC voltage source that converts chemical energy into electrical energy.

T F 5. Diamagnetic materials are only repelled or attracted by a magnetic field.

T F 6. Most pieces of iron are not magnets even though iron is ferromagnetic.

T F 7. Thermoelectricity is heat produced by electrical energy.

T F 8. Magnetism is the most common source of electrical energy.

T F 9. A generator is used to turn electrical energy into mechanical energy.

T F 10. Magnetic lines of force can be harmful to equipment.

T F 11. Over half of all lightning flashes occur within a single cloud.

T F 12. Friction is seldom used to produce electrical energy.

T F 13. Some magnets have only one pole.

T F 14. A lightning rod system prevents lightning strokes by controlling the path of the strokes.

T F 15. An airplane may accumulate a static charge in flight that could interfere with radio communications.

T F 16. Photovoltaic cells cannot be used in watches or calculators.

T F 17. Some thermocouples can measure temperatures as high as 1800°C.

Multiple Choice

______________ **1.** An electrostatic precipitator is a device that uses electricity to remove particles from ___.
- A. lightning strokes
- B. flue gases
- C. photovoltaic systems
- D. cathode-ray tubes

______________ **2.** The electrical polarization of some materials when mechanically strained is called the ___ effect.
- A. photovoltaic
- B. static electricity
- C. magnetic
- D. piezoelectric

______________ **3.** A magnetic ___ is invisible and is produced by a current-carrying conductor, a permanent magnet, or the earth.
- A. field
- B. pole
- C. voltage
- D. process

______________ **4.** The ___ on a motor shaft is used to produce work.
- A. handle
- B. torque
- C. meter
- D. current

______________ **5.** ___ is not a paramagnetic material.
- A. Aluminum
- B. Platinum
- C. Sodium
- D. Chromium

______________ **6.** Cathode-ray tubes are used in ___.
- A. televisions
- B. oscilloscopes
- C. computer monitors
- D. all of the above

______________ **7.** ___ is the resistance to motion that occurs when two surfaces move against each other.
- A. Friction
- B. Static
- C. Reluctance
- D. Magnetism

_______________ **8.** The following substances would all be classified as nonmagnetic except ___.

A. wood
B. glass
C. aluminum
D. water

_______________ **9.** Photoconductive cells are commonly used as sensors in all the following except ___.

A. streetlights
B. smoke detectors
C. phones
D. automatic doors

_______________ **10.** Another name for the Seebeck effect is the ___ effect.

A. photoelectric
B. thermoelectric
C. photovoltaic
D. chemical

_______________ **11.** Photovoltaic cells can be connected in a ___ circuit.

A. series
B. parallel
C. series/parallel
D. all of the above

_______________ **12.** Thermocouples are useful for measuring ___.

A. temperature
B. electron output
C. light energy
D. weight

_______________ **13.** Ionization is the separation of atoms and molecules into particles that have ___ charges.

A. hydraulic
B. chemical
C. electrical
D. mechanical

_______________ **14.** Artificial magnets are often produced because natural magnets have a small ___.

A. shape
B. supply
C. voltage
D. force

______ **15.** A cathode-ray tube is an application of ___.
- A. static electricity
- B. thermionic emission
- C. a magnetic field
- D. chemical action

______ **16.** ___ transducers are used in microphones, timing circuits, and television receivers.
- A. Photovoltaic
- B. Thermionic
- C. Piezoelectric
- D. Magnetic

______ **17.** Thermionic emission is based on ___ energy.
- A. static
- B. electrical
- C. heat
- D. magnetic

______ **18.** A(n) ___ chemically stores and releases its electrical potential when a load is placed across its electrodes.
- A. battery
- B. switch
- C. scope
- D. atom

Completion

______ **1.** A photovoltaic cell converts light energy into ___ energy.

______ **2.** Magnets attract ___ to produce a magnetic field.

______ **3.** ___ is the opposition a substance presents to the passage of magnetic lines of force.

______ **4.** A lightning ___ is the initial electrostatic discharge and return discharge.

______ **5.** Every magnet must have at least two ___ defined as north and south.

______ **6.** A(n) ___ is the unit used as an expression of magnetic force.

______ **7.** Flux ___ is the measure of the magnetic lines of force per unit area.

______ **8.** A lightning rod system ___ a charge from a lightning stroke by providing a conductive path for electrons.

______ **9.** Photoconductive cells are also known as ___.

_______________ **10.** A(n) ___ is a semiconductor device that emits light when an electrical current is present.

_______________ **11.** A motor converts ___ energy into ___ energy.

_______________ **12.** The Seebeck effect uses a(n) ___ difference between two electrical conductors to produce a voltage potential.

_______________ **13.** The ___ effect refers to the absorption or emission of heat at the junctions of two dissimilar metals.

_______________ **14.** An ionic balance is achieved when negative and positive charges are ___.

_______________ **15.** In a photovoltaic cell, ___ flow from the negative terminal to the positive terminal.

_______________ **16.** ___ action was the first reliable and usable method devised for producing electrical energy.

Identification—Effects of Heat

_______________ **1.** Peltier

_______________ **2.** Thermionic

_______________ **3.** Seebeck

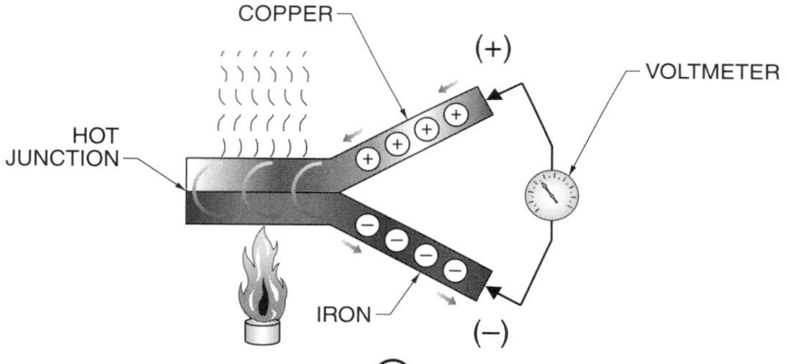

COPPER
(+)
VOLTMETER
HOT JUNCTION
IRON
(–)
Ⓐ

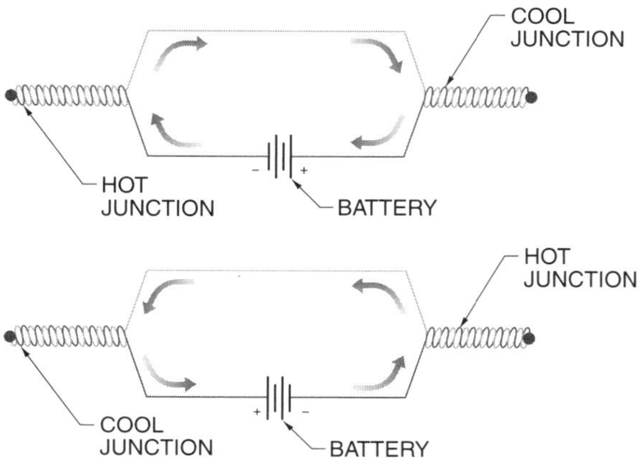

COOL JUNCTION
HOT JUNCTION
BATTERY
HOT JUNCTION
COOL JUNCTION
BATTERY
Ⓑ

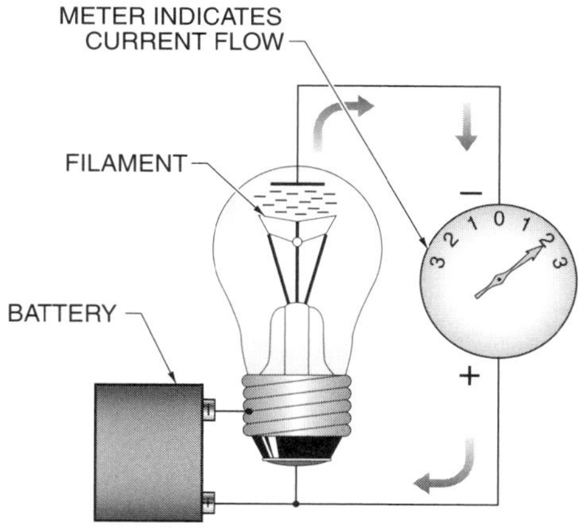

METER INDICATES CURRENT FLOW
FILAMENT
BATTERY
Ⓒ

Thermocouple Voltages

Modern thermocouple calibration instruments automatically adjust to maintain a 32°F (0°C) ice point reference for both reading and simulating thermocouples. Although this is automatic, it is important to understand the basis for these automatic operations. The values shown are in millivolts referenced to the 32°F (0°C) mV value. The voltage produced at the ambient temperature must be added to the voltage produced at the elevated temperature.

Type J Thermoelectric Voltage

°C	0	–1	–2	–3	–4	–5	–6	–7	–8	–9	–10
–40	–1.961	–2.008	–2.055	–2.103	–2.150	–2.197	–2.244	–2.291	–2.338	–2.385	–2.431
–30	–1.482	–1.530	–1.578	–1.626	–1.674	–1.722	–1.770	–1.818	–1.865	–1.913	–1.961
–20	–0.995	–1.044	–1.093	–1.142	–1.190	–1.239	–1.288	–1.336	–1.385	–1.433	–1.482
–10	–0.501	–0.550	–0.600	–0.650	–0.699	–0.749	–0.798	–0.847	–0.896	–0.946	–0.995
0	0.000	–0.050	–0.101	–0.151	–0.201	–0.251	–0.301	–0.351	–0.401	–0.451	–0.501

°C	0	1	2	3	4	5	6	7	8	9	10
0	0.000	0.050	0.101	0.151	0.202	0.253	0.303	0.354	0.405	0.456	0.507
10	0.507	0.558	0.609	0.660	0.711	0.762	0.814	0.865	0.916	0.968	1.019
20	1.019	1.071	1.122	1.174	1.226	1.277	1.329	1.381	1.433	1.485	1.537
30	1.537	1.589	1.641	1.693	1.745	1.797	1.849	1.902	1.954	2.006	2.059
40	2.059	2.111	2.164	2.216	2.269	2.322	2.374	2.427	2.480	2.532	2.585
50	2.585	2.638	2.691	2.744	2.797	2.850	2.903	2.956	3.009	3.062	3.116
60	3.116	3.169	3.222	3.275	3.329	3.382	3.436	3.489	3.543	3.596	3.650
70	3.650	3.703	3.757	3.810	3.864	3.918	3.971	4.025	4.079	4.133	4.187
80	4.187	4.240	4.294	4.348	4.402	4.456	4.510	4.564	4.618	4.672	4.726
90	4.726	4.781	4.835	4.889	4.943	4.997	5.052	5.106	5.160	5.215	5.269

in mV

The following millivolt (mV) readings were made with a digital voltmeter with a Type J thermocouple and no ice point compensation. The ambient temperature is provided. Determine the actual temperature at the hot junction.

__________ **1.** Given an ambient temperature of 0°C and a voltage reading of +1.277 mV, the hot junction temperature is ____.

__________ **2.** Given an ambient temperature of 15°C and a voltage reading of +4.181 mV, the hot junction temperature is ____.

__________ **3.** Given an ambient temperature of 20°C and a voltage reading of –0.154 mV, the hot junction temperature is ____.

__________ **4.** Given an ambient temperature of 25°C and a voltage reading of –2.173 mV, the hot junction temperature is ____.

__________ **5.** Given an ambient temperature of –10°C and a voltage reading of +5.336 mV, the hot junction temperature is ____.

The Simple Circuit and Ohm's Law

Name ______________________________ Date ______________

True-False

T F **1.** To show that conductors are connected, a dot may be placed at their intersection.

T F **2.** In order to measure current flow through a component, a meter must be connected in series with the component.

T F **3.** Lamps, speakers, and motors are all considered common loads.

T F **4.** Meter loading refers to inaccurate readings due to the meter's resistance being in parallel with the component being tested.

T F **5.** In DC circuits, the higher the resistance, the higher the current flow.

T F **6.** Overloads can be caused by defective circuit components.

T F **7.** Switches can only be activated mechanically.

T F **8.** Power can be defined as the rate of doing work.

T F **9.** A three-way switch has one common terminal and two traveler terminals.

T F **10.** A short circuit occurs when a circuit has a higher resistance than the normal circuit resistance.

T F **11.** Switches are used as control devices in a circuit.

T F **12.** Power is distributed in a circuit that has voltage but no electron flow.

T F **13.** A speaker can convert electrical energy into sound waves.

T F **14.** A partial or dead short can result in a fire.

T F **15.** Four-way switches have ON and OFF markings on them.

T F **16.** A plug fuse has a screw-in base similar to a light bulb.

Multiple Choice

_______________ **1.** Most switches are rated based on the ___ current and ___ voltage they can safely handle.
A. maximum; minimum
B. maximum; maximum
C. minimum; maximum
D. minimum; minimum

_______________ **2.** Power supplied by a circuit is ___ the power consumed by a circuit.
A. equal to
B. greater than
C. less than
D. not related to

_______________ **3.** Clamp-on ammeters measure ___ in a circuit.
A. temperature
B. magnetism
C. electron flow
D. current

_______________ **4.** A ___ switch is mechanically operated.
A. pushbutton
B. selector
C. limit
D. foot

_______________ **5.** A simple electrical circuit must have a(n) ___, a voltage source, and a load.
A. conductor
B. motor
C. inductor
D. relief valve

_______________ **6.** A(n) ___ circuit has no electron flow.
A. closed
B. complex
C. open
D. large

_______________ **7.** A ___ is the part of a switch that determines the number of circuits the switch can control.
A. lamp
B. pole
C. break
D. throw

_______________ **8.** A(n) ___ is an example of an overcurrent protection device.
A. circuit breaker
B. electrical cord
C. generator
D. switch

_______________ **9.** In Ohm's law, the voltage of a circuit divided by the resistance is equal to the ___.
A. speed
B. current
C. flux
D. conductivity

_______________ **10.** ___ switches need little or no input to operate.
A. Manual
B. Insulating
C. Heat
D. Automatic

_______________ **11.** An electric ___ is a rotating device that converts electrical power into a rotating mechanical force.
A. motor
B. valve
C. meter
D. magnet

_______________ **12.** The interrupting rating is related to the amount of ___ that a fuse can safely suspend without arcing over.
A. voltage
B. resistance
C. current
D. power

_______________ **13.** ___ devices are required in motor circuits by the National Electrical Code®.
A. Overload
B. Magnetism
C. Phase shift
D. Motion sensor

_______________ **14.** When possible, conductors being tested with a clamp-on ammeter should be ___ surrounding conductors.
A. next to
B. separated from
C. on top of
D. below

Completion

______________ **1.** The ___ method is the preferred way of showing that conductors are not connected.

______________ **2.** Electrical measurements are commonly taken using a(n) ___.

______________ **3.** A(n) ___ occurs when circuit current rises above the level at which the load and/or circuit is designed to operate.

______________ **4.** When a substance is heated to a high temperature, the resulting emission of light energy is referred to as ___.

______________ **5.** ___ is energy that consists of pressure vibrations in the air.

______________ **6.** Ohm's law is the relationship between voltage, current, and ___ in an electrical circuit.

______________ **7.** A conductor is used to carry electric ___.

______________ **8.** The power formula is often referred to as ___ law.

______________ **9.** A(n) ___ is a place on a switch contact that opens or closes a circuit.

______________ **10.** A(n) ___ switch has one or more poles and can be connected in several positions.

______________ **11.** The voltage rating of a(n) ___ must be equal to or greater than that of the voltage source in the circuit.

______________ **12.** A(n) ___ will manually or automatically open a circuit when a short circuit or overload occurs.

______________ **13.** A(n) ___ is a safe device for operators to use because it uses only a small input current to open or close contacts in a load circuit.

______________ **14.** The power formula and ___ law can be combined to find any unknown value if two of the values are known.

Identification—Switch Contacts

______________ **1.** Normally open

______________ **2.** Single-pole, double-throw, double-break

______________ **3.** Single-pole, single-throw, single-break

______________ **4.** Normally closed

______________ **5.** Single-pole, single-throw, double-break

______________ **6.** Multiple-contact switch

______________ **7.** Double-pole, double-throw, double-break

______________ **8.** Double-pole, single-throw, single-break

______________ **9.** Single-pole, double-throw, single-break

CLOSED CONTACT
(A)

MECHANICALLY CONNECTED
(B)

MECHANICALLY CONNECTED
(C)

COMMON TERMINAL
(D)

(E)

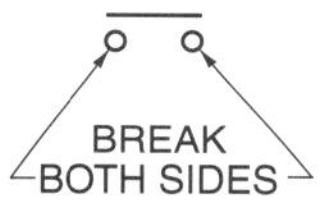

BREAK BOTH SIDES
(F)

POLE
TERMINALS
(G)

CONTACT CLOSED
SOLID-STATE CONTACTS
(H)

MECHANICAL CONTACTS
SOLID-STATE CONTACTS
(I)

Identification—Switches

________________ **1.** Pushbuttons

________________ **2.** Foot switches

________________ **3.** Selector switches

________________ **4.** Level switches

________________ **5.** Temperature switches

________________ **6.** Flow switches

________________ **7.** Proximity switches

________________ **8.** Limit switches

________________ **9.** Pressure switches

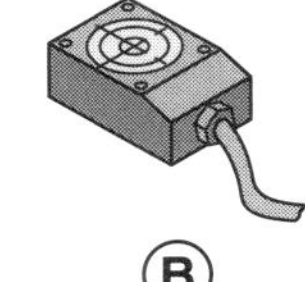

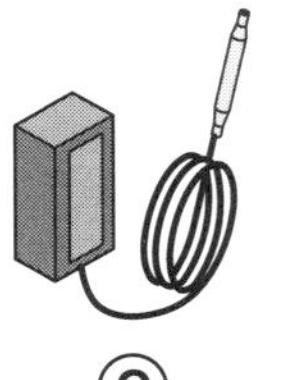

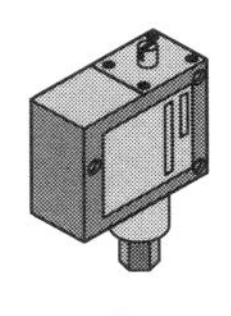

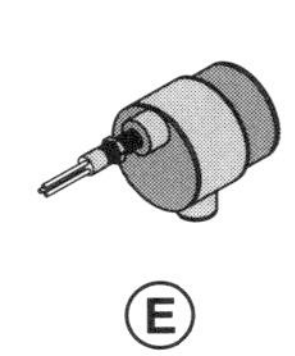

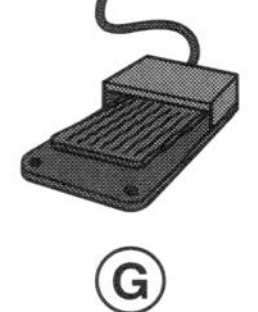

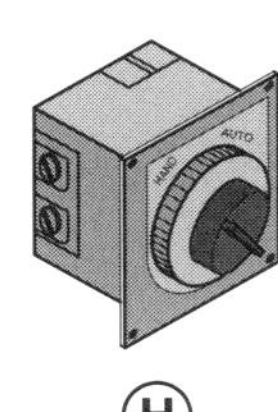

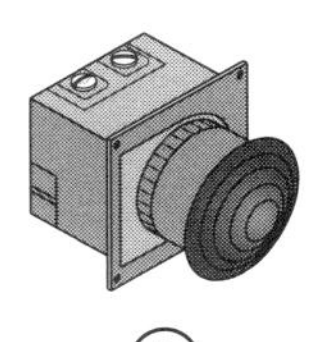

Calculating Current

Calculate the current for each value of voltage.

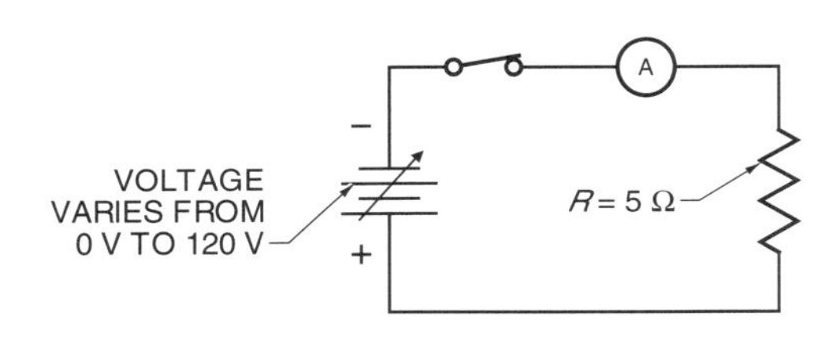

________________ **1.** $V = 0$ V; $I =$ ___ A

________________ **2.** $V = 6$ V; $I =$ ___ A

________________ **3.** $V = 12$ V; $I =$ ___ A

________________ **4.** $V = 24$ V; $I =$ ___ A

________________ **5.** $V = 48$ V; $I =$ ___ A

________________ **6.** $V = 120$ V; $I =$ ___ A

Calculating Voltage

Calculate the voltage for each value of current.

_______________ **1.** I = 0 A; V = ___ V

_______________ **2.** I = 2 A; V = ___ V

_______________ **3.** I = 3 A; V = ___ V

_______________ **4.** I = 6 A; V = ___ V

_______________ **5.** I = 10 A; V = ___ V

_______________ **6.** I = 12 A; V = ___ V

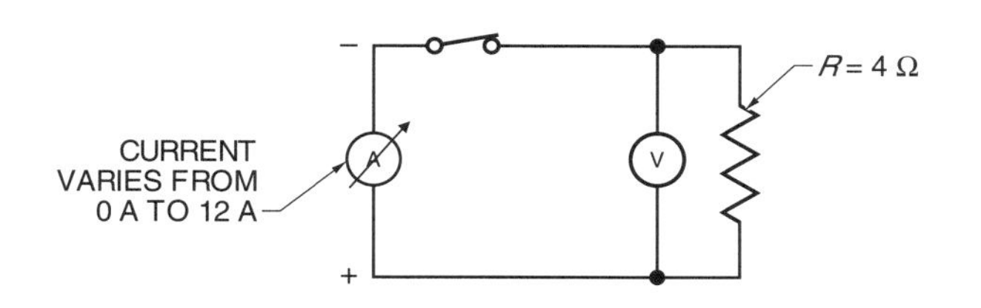

Calculating Power

Calculate the power for each value of voltage and current.

_______________ **1.** V = 0 V; I = 0 A; P = ___ W

_______________ **2.** V = 6 V; I = 0.25 A; P = ___ W

_______________ **3.** V = 12 V; I = 0.375 A; P = ___ W

_______________ **4.** V = 24 V; I = 0.50 A; P = ___ W

_______________ **5.** V = 48 V; I = 0.75 A; P = ___ W

_______________ **6.** V = 120 V; I = 1 A; P = ___ W

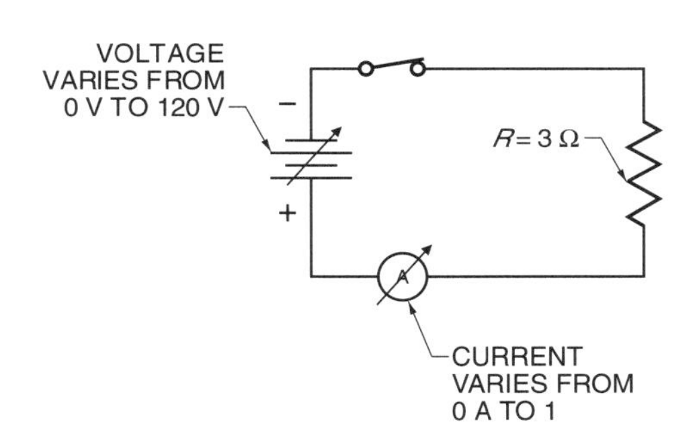

Light Circuits

____________________ **1.** P_T = ___ W

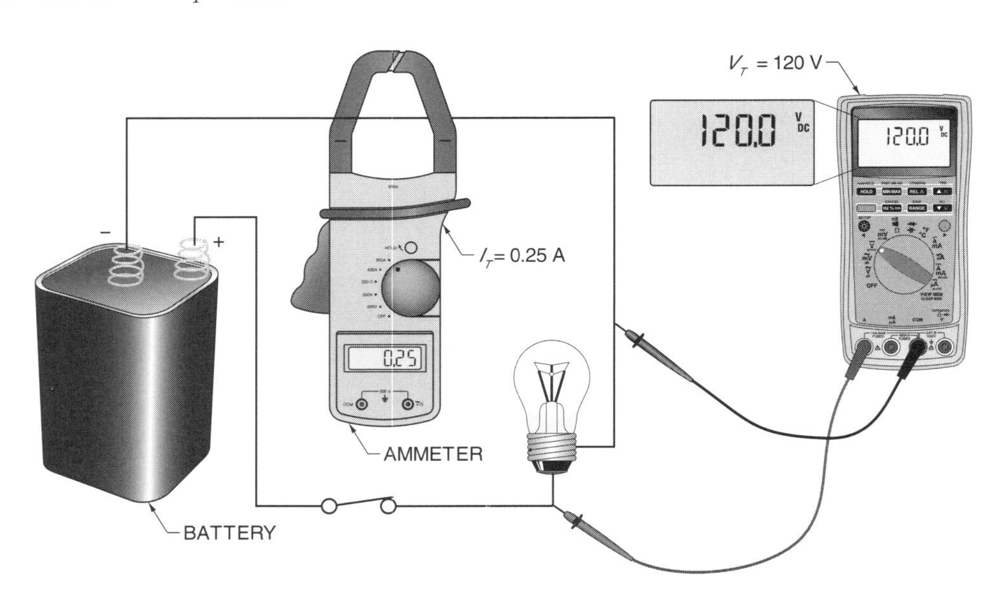

____________________ **2.** V_T = ___ V

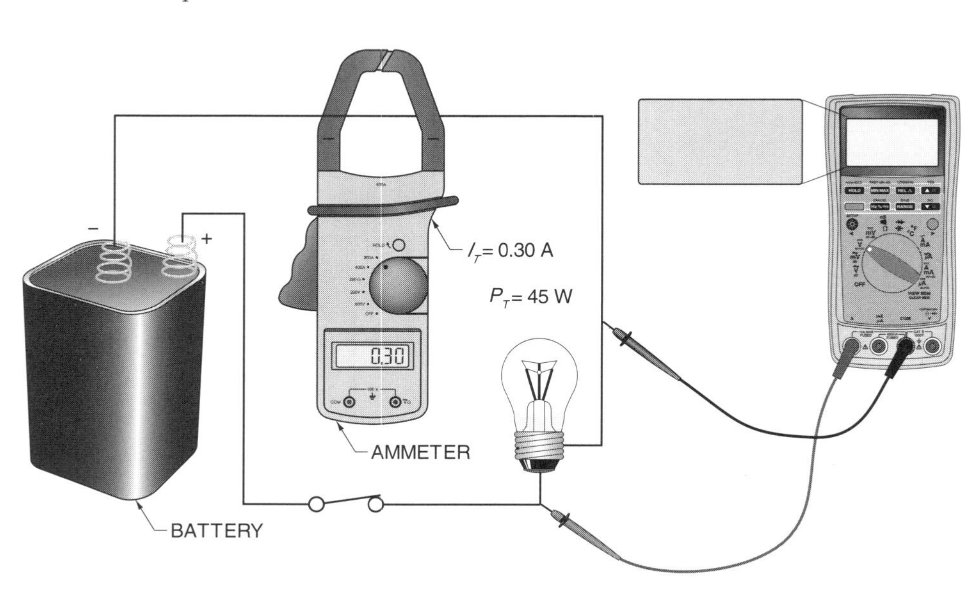

____________________ **3.** I_T = ___ A

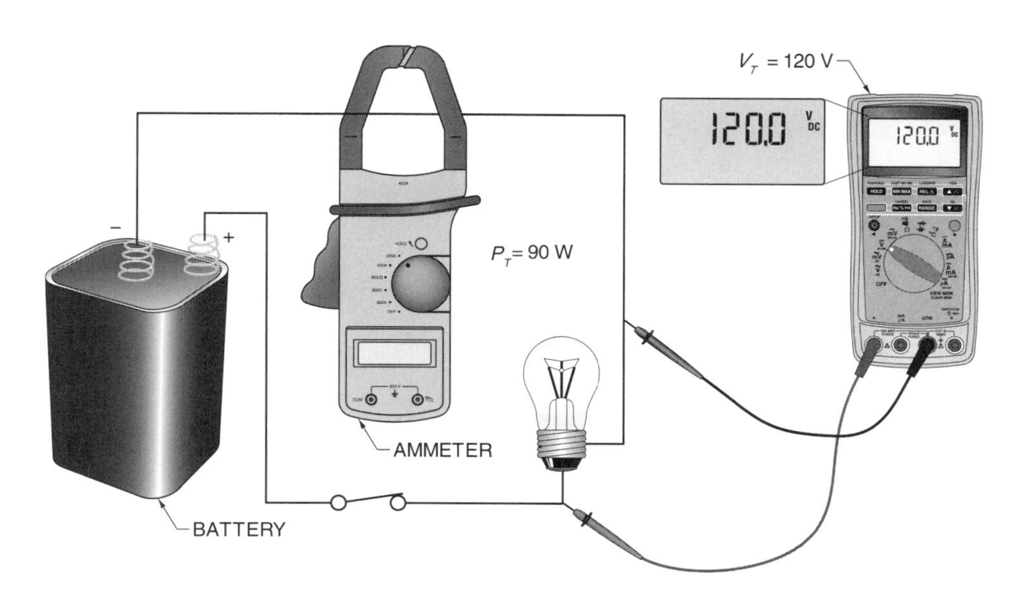

DC Series Circuits 5

Name ______________________________ Date ______________

True-False

T F **1.** In a series circuit, there are three paths for electron flow.

T F **2.** The lower a load's resistance, the more power produced in a series circuit.

T F **3.** Heat is produced when electrons flow through a wire that has resistance.

T F **4.** A voltage multiplier is a circuit made of resistors connected in series to produce a desired voltage drop across the resistors.

T F **5.** Some electrical sources have no internal resistance.

T F **6.** Efficiency of a circuit can be determined by dividing the power at the load by the total power output.

T F **7.** In a series circuit, the sum of the voltage increases is equal to the source voltage.

T F **8.** A complex circuit includes two or more resistive loads.

T F **9.** Electrons have bodies that can be counted.

T F **10.** Opening a series circuit at any point will increase the flow of electrons.

T F **11.** Assuming that power at each load in a series circuit is known, total power can be calculated by adding the power at each load.

T F **12.** Resistors are available with a variety of power ratings.

Multiple Choice

______________ **1.** ___ is the flow of electrons through an electrical circuit.
A. Series
B. Parallel
C. Current
D. Resistance

______________ **2.** Voltage ___ is the amount of voltage used by a device when current passes through it.
A. drop
B. rise
C. quantity
D. quality

______________ **3.** A ground is an unlimited source and acceptor of ___ in a circuit.
A. neutrons
B. electrons
C. power
D. polarity

______________ **4.** Maximum ___ transfer between a source and load depends on the load resistance matching the source resistance.
A. electron
B. power
C. voltage
D. speed

______________ **5.** To find total resistance in a series circuit, the values of the resistors in the circuit are ___.
A. subtracted
B. multiplied
C. divided
D. added

Completion

______________ **1.** Current has the same value at any point in a(n) ___ circuit.

______________ **2.** ___ produced is measured in watts (W).

______________ **3.** ___ value determines variables such as wire size and fuse rating in an electrical system.

______________ **4.** Ohm's law can be used to calculate the ___ drop across any resistor.

______________ **5.** A power rating is listed in watts for appliances and in ___ for motors.

Calculating Total Resistance and Total Current

__________________ **1.** R_T = ___ Ω

__________________ **2.** I_T = ___ A

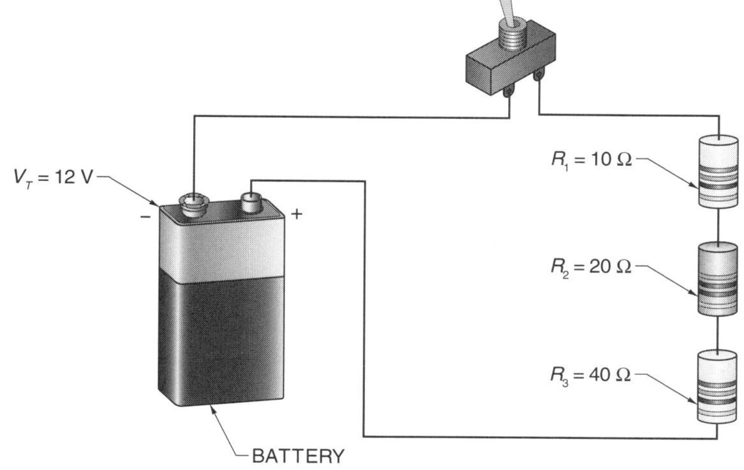

__________________ **3.** R_T = ___ Ω

__________________ **4.** I_T = ___ A

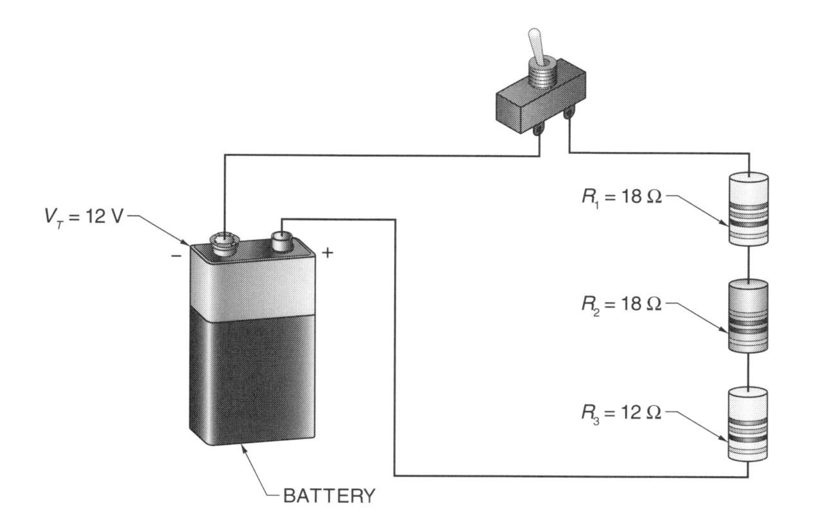

_______________ **5.** R_T = ___ Ω

_______________ **6.** I_T = ___ A

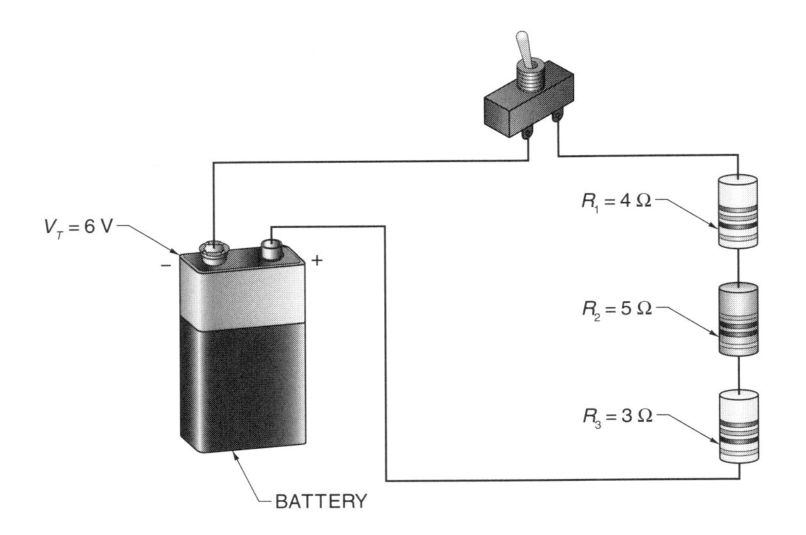

_______________ **7.** R_T = ___ Ω

_______________ **8.** I_T = ___ A

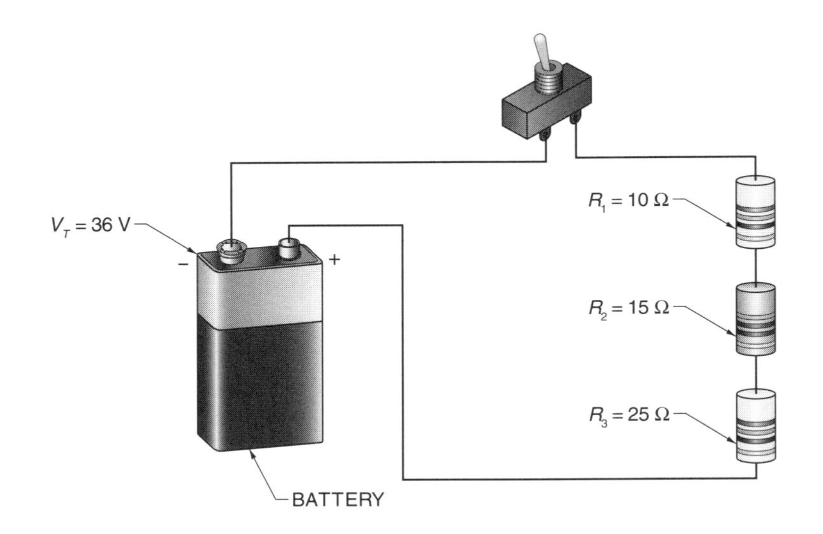

Calculating Voltage Drop

____________ **1.** V_1 = ___ V

____________ **2.** V_2 = ___ V

____________ **3.** V_T = ___ V

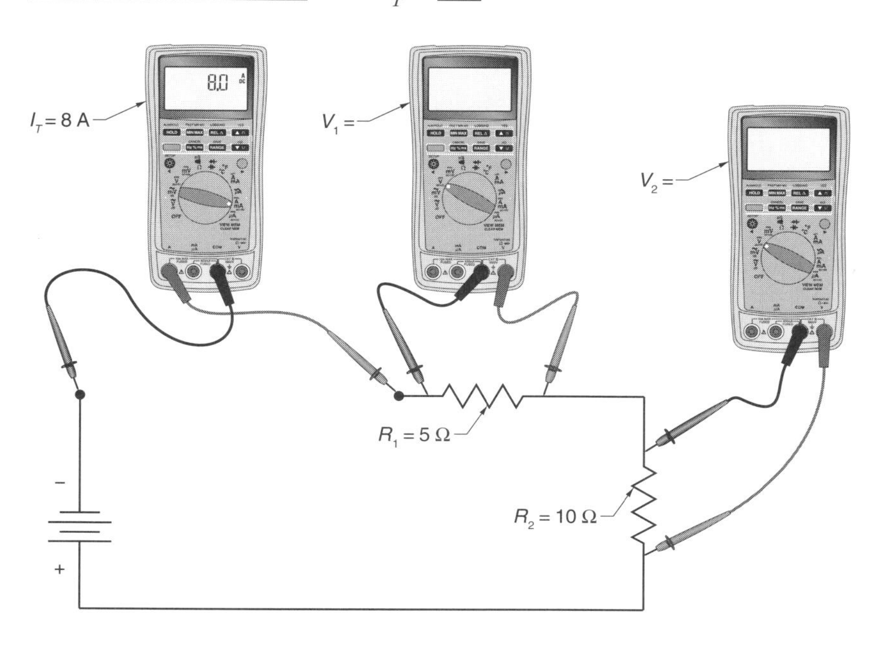

____________ **4.** V_1 = ___ V

____________ **5.** V_2 = ___ V

____________ **6.** V_T = ___ V

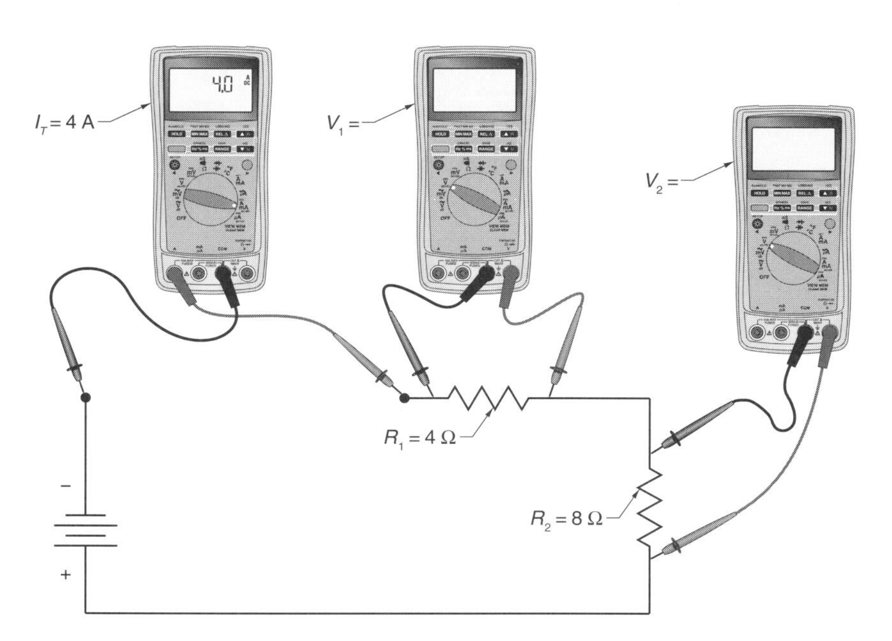

______________ **7.** V_1 = ___ V

______________ **8.** V_2 = ___ V

______________ **9.** V_T = ___ V

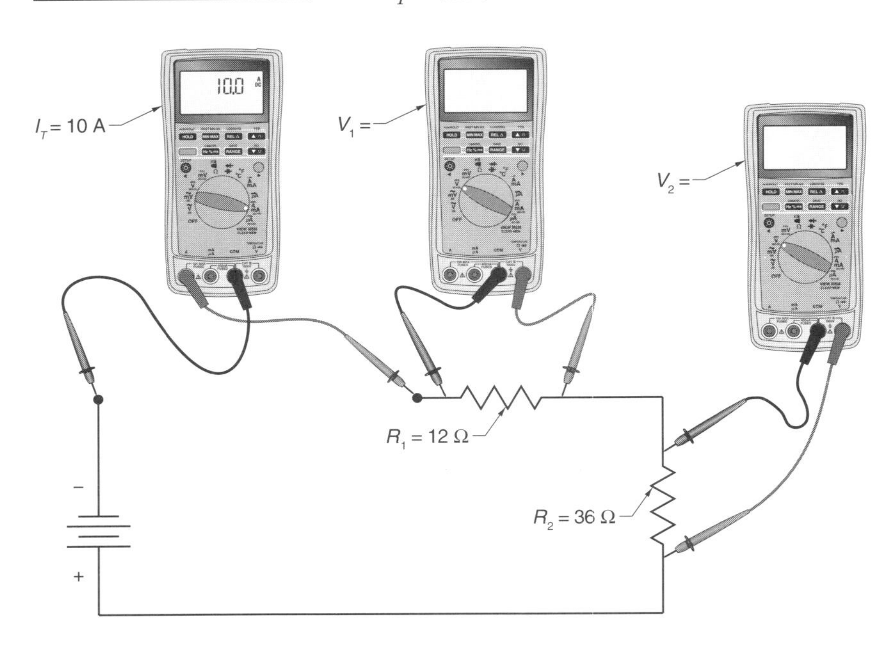

______________ **10.** V_1 = ___ V

______________ **11.** V_2 = ___ V

______________ **12.** V_T = ___ V

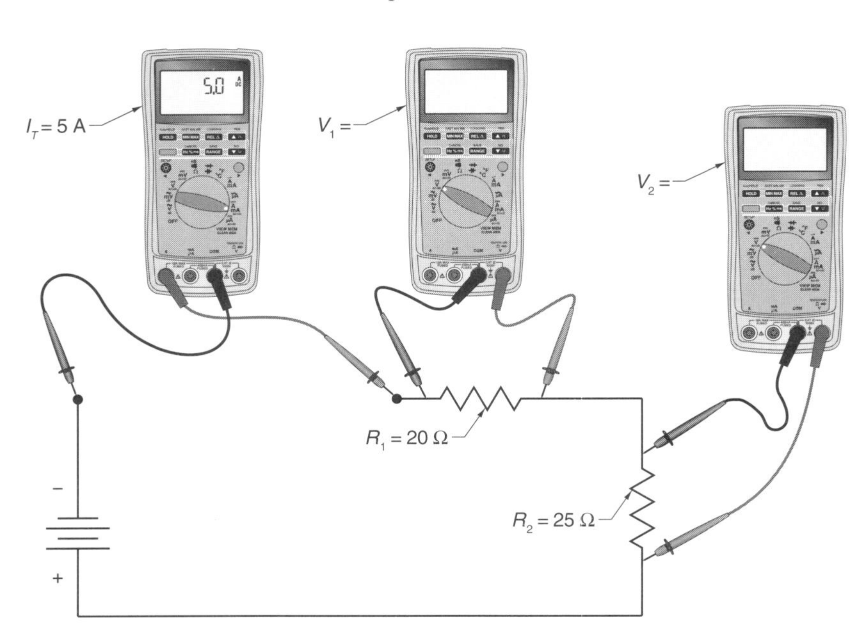

Coffeemaker Circuit

A typical coffeemaker demonstrates how a basic series circuit is used to brew coffee and keep it warm. The brew heating element heats the water and forces it over the coffee grounds. The warm heating element keeps the coffee warm after brewing. When the brew cycle begins, the only resistance in the circuit is that of the brew heating element. The warm heating element is not part of the circuit at this time.

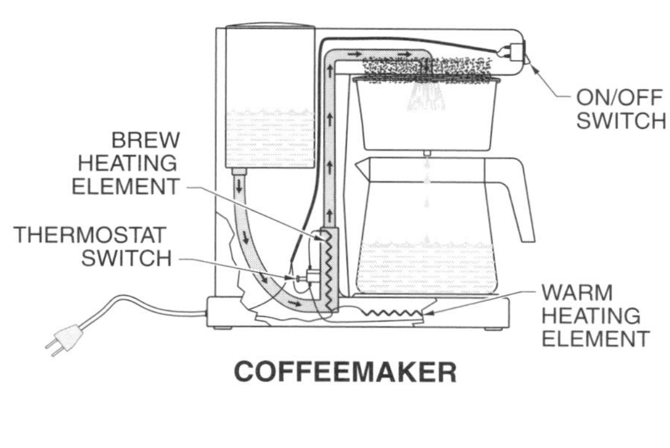

COFFEEMAKER

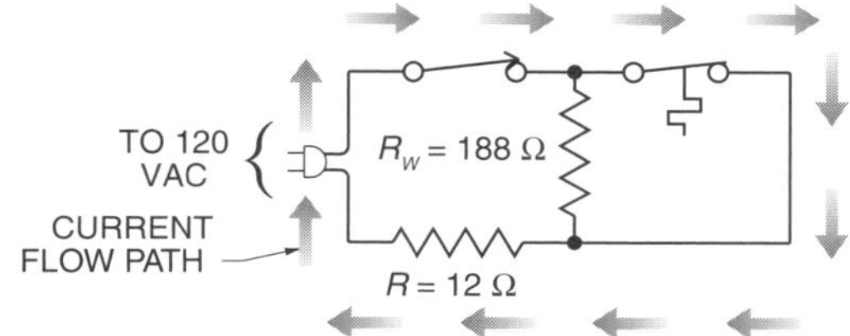

BREWING CIRCUIT CONDITION

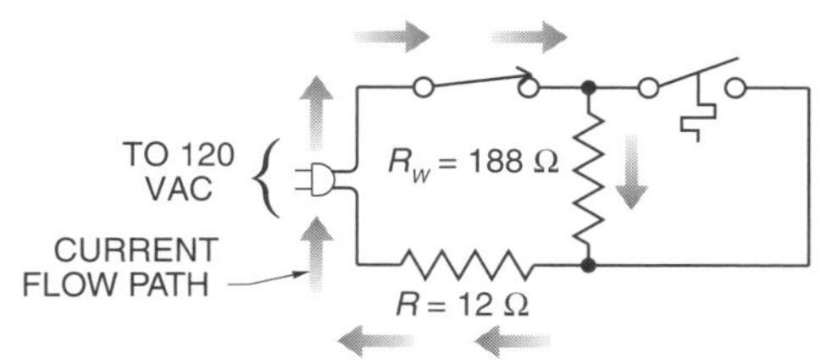

WARMING CIRCUIT CONDITION

Using the Brewing Circuit Condition figure, calculate the brewing current and power.

______________________ **1.** I_B = ___ A

______________________ **2.** P_B = ___ A

After all the water is forced over the brew heating element, the temperature increases and opens the switch. The warm heating element and the brew heating element are connected in series when this occurs.

Using the Warming Circuit Condition figure, calculate the total resistance, warming current, and warming power.

________________ **3.** R_T = ___ Ω

________________ **4.** I_W = ___ A

________________ **5.** P_W = ___ A

Voltage Dividers

Voltage dividers are often used to adjust voltage in televisions and radios. In a voltage divider, the full voltage is applied across the total resistance, with a portion of the full voltage available across each resistor.

Calculate the voltage across each section of the voltage divider.

________________ **1.** V_1 = ___ V

________________ **2.** V_2 = ___ V

________________ **3.** V_3 = ___ V

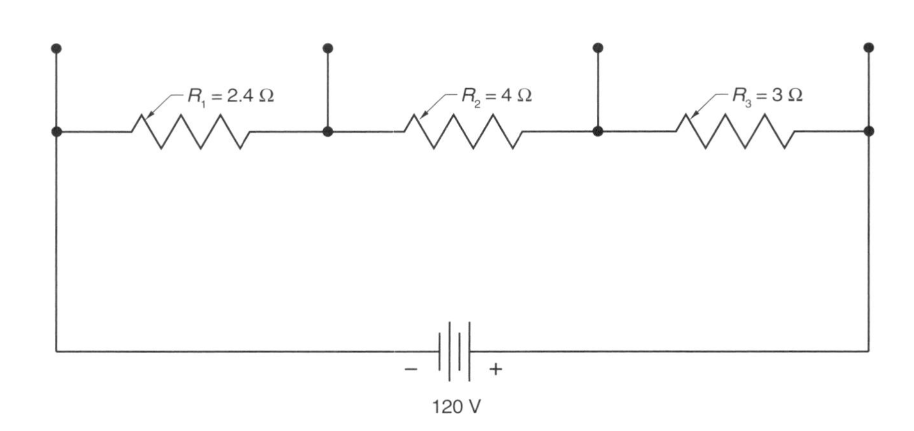

DC Parallel Circuits 6

Name ______________________________ Date ______________

True-False

T F **1.** In a DC parallel circuit, electrons flow in one direction.

T F **2.** Conductance is improved and resistance reduced when adding resistors in parallel.

T F **3.** As loads are added in parallel, the total current decreases.

T F **4.** Connecting loads in parallel forces them to operate at different voltage levels.

T F **5.** As the resistance of a load decreases, the amount of power produced increases.

T F **6.** An open circuit is the same as a series circuit in that any open point in the circuit causes electron flow to stop.

T F **7.** When current takes a shortcut around the normal path of current flow, it is referred to as a short circuit.

T F **8.** A short-circuited branch circuit has a resistance of zero.

T F **9.** Total current in a parallel circuit is calculated by adding the current in all the branches.

T F **10.** Ohm's law can be used to calculate total circuit voltage and current but not total circuit resistance.

Multiple Choice

______________ **1.** The voltage drop in a parallel circuit ___ across each load.

A. increases
B. approaches zero
C. stays the same
D. decreases

______________ **2.** Zero current flow may indicate an open between the ___ and a load or switch in the circuit.
A. power source
B. resistor
C. branch
D. insulation

______________ **3.** In a parallel circuit, each load is called a(n) ___ circuit.
A. branch
B. short
C. voltage drop
D. ohm

______________ **4.** Doubling the number of conductors in parallel will ___ the electron flow in a circuit.
A. triple
B. half
C. double
D. not affect

______________ **5.** The total resistance of the resistors in a branch circuit is calculated by using the ___ method.
A. addition
B. reciprocal
C. integral
D. inverse

Completion

______________ **1.** A(n) ___ circuit has two or more loads and more than one path for electron flow.

______________ **2.** A(n) ___ resistor may be added to a parallel circuit to produce a constant flow of current.

______________ **3.** Electron flow is interrupted in a(n) ___ circuit.

______________ **4.** Total power of a parallel circuit is equal to the ___ of the power produced by each load.

______________ **5.** Total voltage multiplied by total circuit current yields total ___ in a circuit.

Voltage Calculations

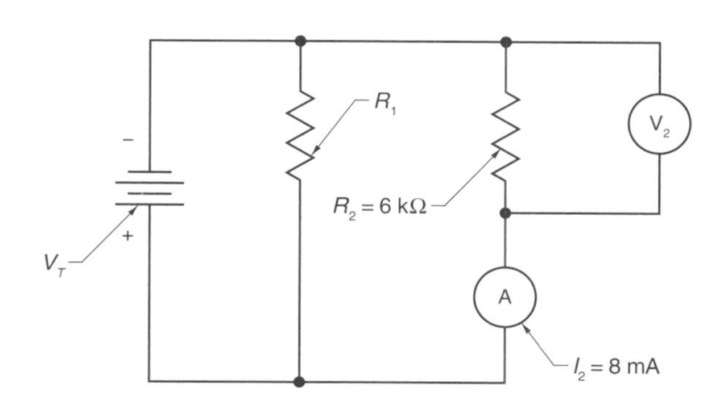

________________ 1. V_2 = ___ V

________________ 2. V_T = ___ V

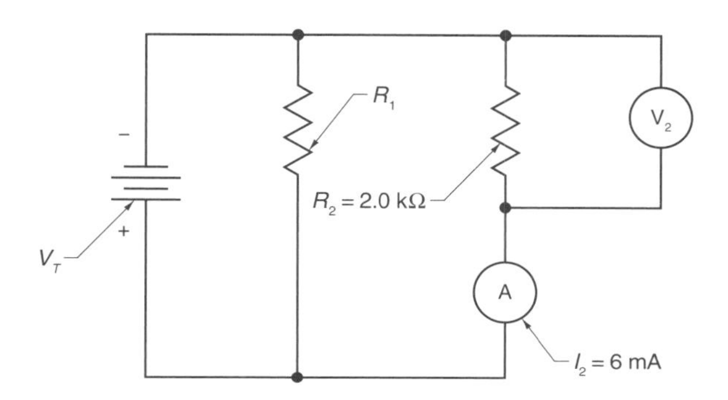

________________ 3. V_2 = ___ V

________________ 4. V_T = ___ V

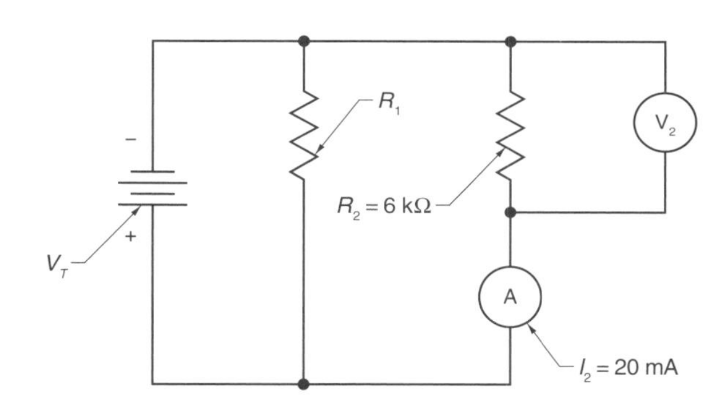

________________ 5. V_2 = ___ V

________________ 6. V_T = ___ V

Current Calculations

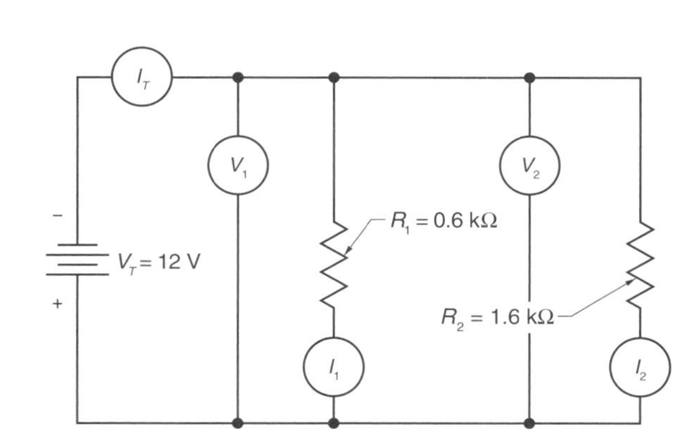

________________ 1. I_1 = ___ A

________________ 2. I_2 = ___ A

________________ 3. I_T = ___ A

________________ **4.** I_1 = ___ A

________________ **5.** I_2 = ___ A

________________ **6.** I_T = ___ A

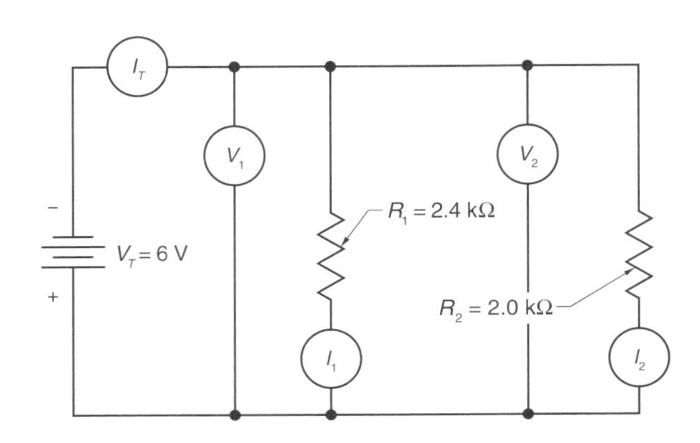

________________ **7.** I_1 = ___ A

________________ **8.** I_2 = ___ A

________________ **9.** I_T = ___ A

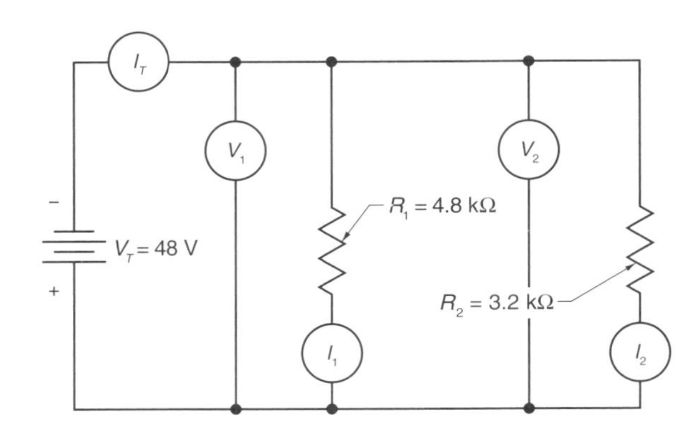

Resistance Calculations

________________ **1.** R_T = ___ Ω

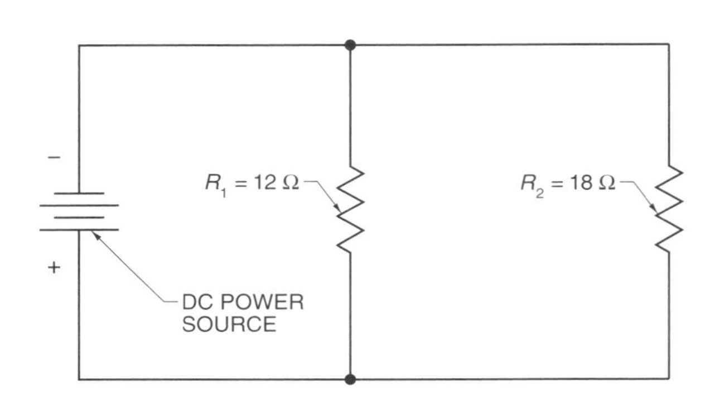

________________ **2.** R_T = ___ Ω

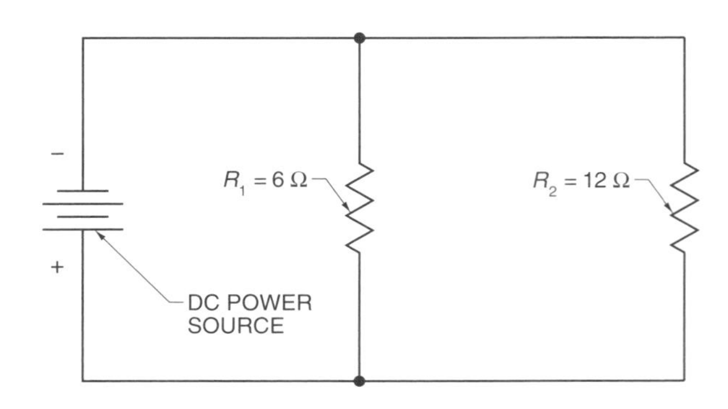

________________ 3. R_T = ___ Ω

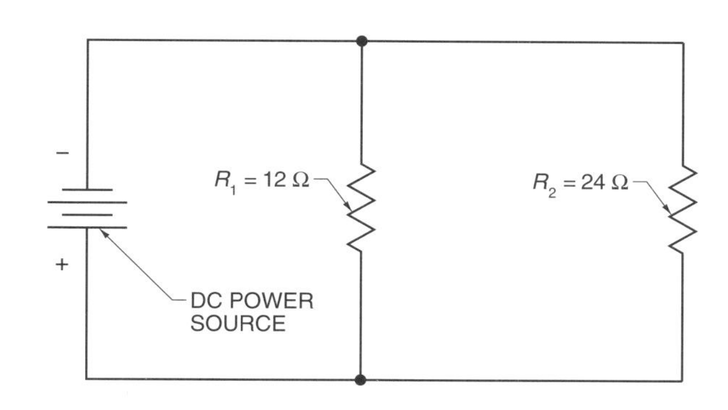

Finding Total Resistance and Current

________________ 1. R_T = ___ Ω

________________ 2. I_1 = ___ A

________________ 3. I_2 = ___ A

________________ 4. I_T = ___ A

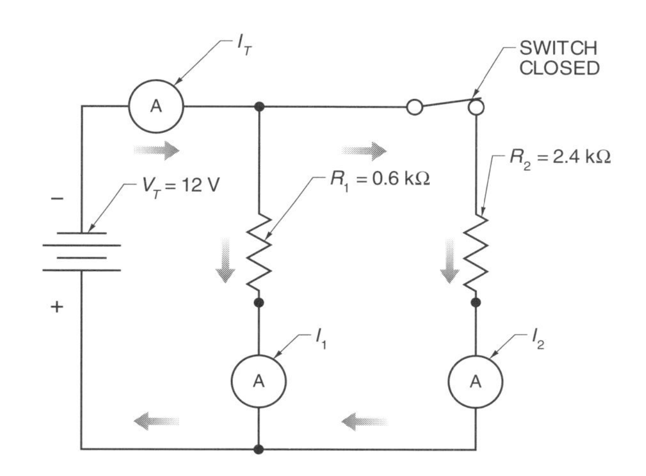

________________ 5. R_T = ___ Ω

________________ 6. I_1 = ___ A

________________ 7. I_2 = ___ A

________________ 8. I_T = ___ A

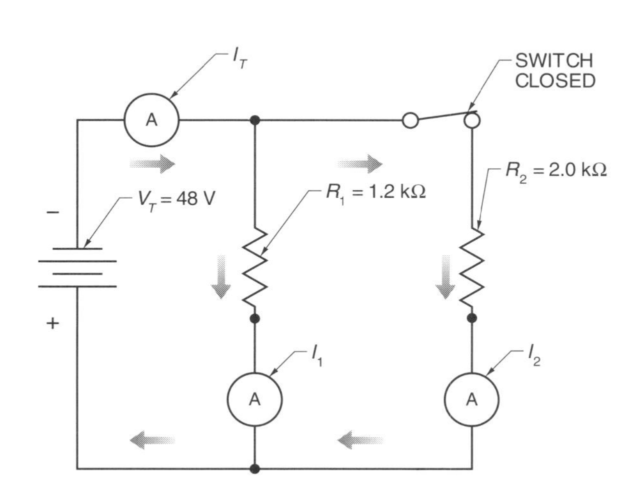

Appliances Connected in Parallel

Calculate the total circuit current of the parallel circuit.

1. I_T = ___ A

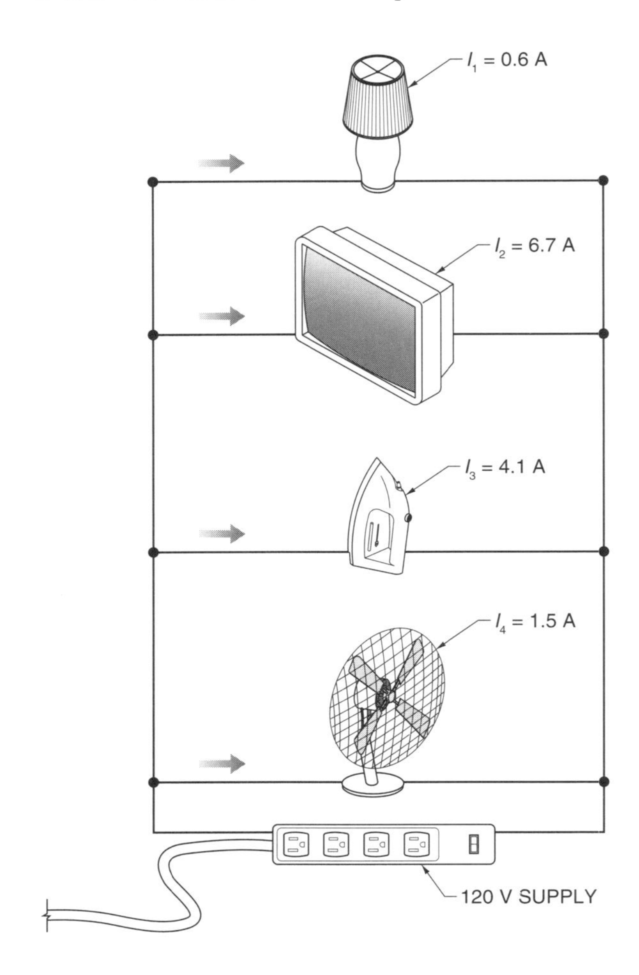

Calculate the total power used in the circuit by first calculating the power used for each appliance.

________________ **2.** P_1 = ___ W

________________ **3.** P_2 = ___ W

________________ **4.** P_3 = ___ W

________________ **5.** P_4 = ___ W

________________ **6.** P_T = ___ W

DC Series/Parallel Circuits 7

Name ______________________________ Date ______________

True-False

T F 1. In a combination circuit, current can flow in part of a parallel circuit while another part of the circuit is turned OFF.

T F 2. Knowing total power and source voltage makes it possible to calculate variables like current.

T F 3. All DC resistance sources have a positive and a negative terminal.

T F 4. A heating element can produce resistance in a series/parallel circuit.

T F 5. Most electronic circuits consist of series-connected components.

T F 6. To solve for unknown variables in a series/parallel circuit, separate calculations are used for the series section and for the parallel section.

T F 7. Switches can be connected only in parallel within a DC circuit.

T F 8. Frequently, the only known variables in a series/parallel circuit are resistance and the source voltage.

T F 9. Total circuit resistance and current must be known before the voltage drop across each resistance of a circuit can be calculated.

T F 10. To do calculations on a series/parallel circuit, the circuit should be broken down into its basic series and parallel parts.

Multiple Choice

______________ 1. A series/parallel circuit can be used to connect a switch and a ___ together.

A. meter
B. fuse
C. circuit breaker
D. all of the above

______________ **2.** A connection point is another term for a ___.
A. positive and negative terminal
B. switch
C. fuse
D. series/parallel circuit

______________ **3.** A(n) ___ resistor is used to increase the range of an ammeter in a parallel system.
A. ballast
B. principal
C. shunt
D. instrument

______________ **4.** Power in a combination circuit is produced when current flows through a(n) ___.
A. ohm
B. resistance
C. voltage
D. conductor

______________ **5.** Total power in a series/parallel circuit is calculated by ___ the power produced by each resistance.
A. adding
B. subtracting
C. multiplying
D. none of the above

Completion

______________ **1.** Series/parallel circuits are often referred to as ___ circuits.

______________ **2.** A(n) ___ is a junction point in a circuit where connections are made between different circuit paths.

______________ **3.** An ammeter is used to measure the amount of ___ in a circuit.

______________ **4.** In order to stop current flow, at least one switch must be ___.

______________ **5.** According to NEC® requirements, circuit conductors cannot have a voltage ___ that is more than 3% of the source voltage.

Calculating Resistance in Series/Parallel Circuits

________________ **1.** R_T = ___ Ω

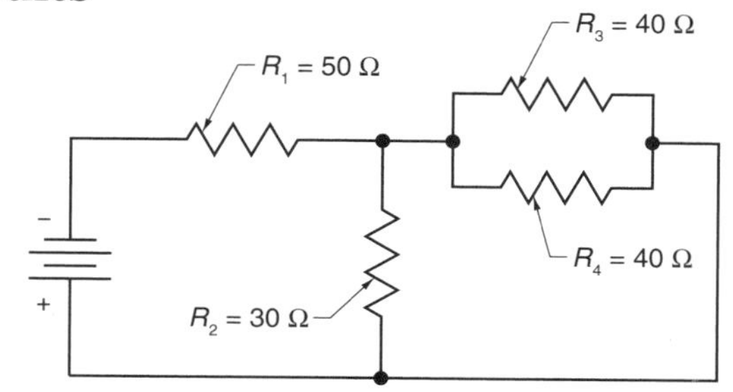

________________ **2.** R_T = ___ Ω

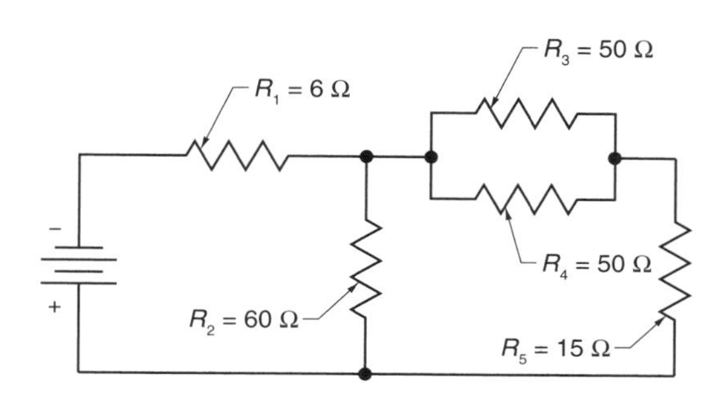

________________ **3.** R_T = ___ Ω

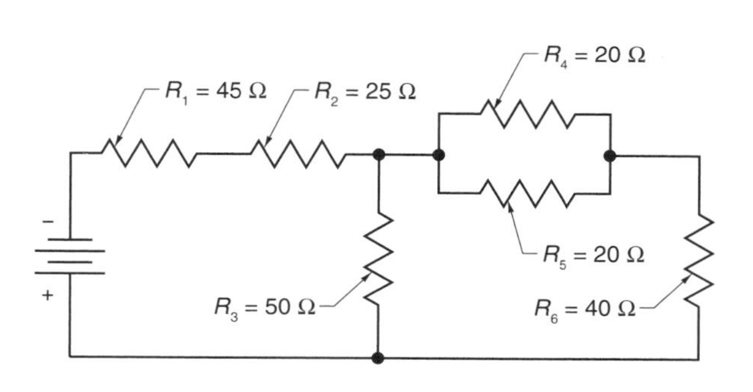

Calculating Voltage Drop in Series/Parallel Circuits

________________ **1.** R_T = ___ Ω

________________ **2.** I_T = ___ A

________________ **3.** V_{S1} = ___ V

________________ **4.** V_{S2} = ___ V

________________ **5.** $V_{P3,4}$ = ___ V

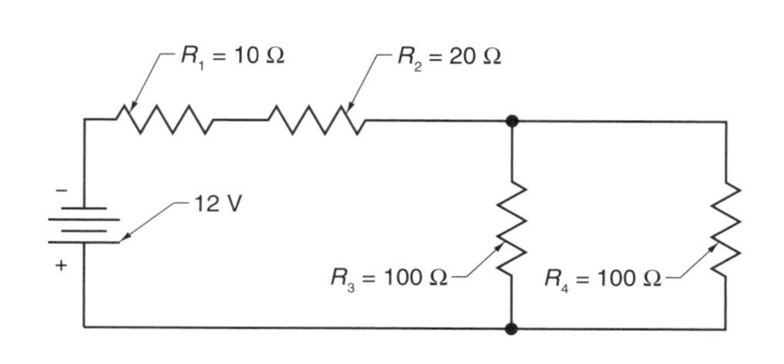

______ **6.** R_T = ___ Ω

______ **7.** I_T = ___ A

______ **8.** V_{S1} = ___ V

______ **9.** V_{S2} = ___ V

______ **10.** $V_{P3,4}$ = ___ V

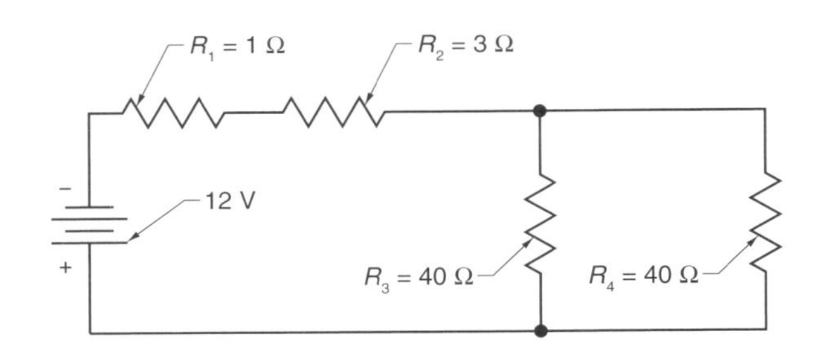

______ **11.** R_T = ___ Ω

______ **12.** I_T = ___ A

______ **13.** V_{S1} = ___ V

______ **14.** V_{S2} = ___ V

______ **15.** $V_{P3,4}$ = ___ V

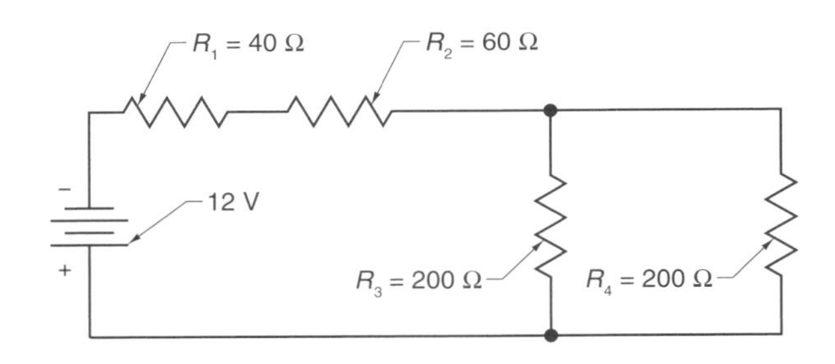

Calculating Total Power in Series/Parallel Circuits

______ **1.** R_T = ___ Ω

______ **2.** I_1 = ___ A

______ **3.** V_D = ___ V

______ **4.** V_P = ___ V

______ **5.** P_{S1} = ___ W

______ **6.** P_{S2} = ___ W

______ **7.** P_{P1} = ___ W

______ **8.** P_{P2} = ___ W

______ **9.** P_{P3} = ___ W

______ **10.** P_T = ___ W

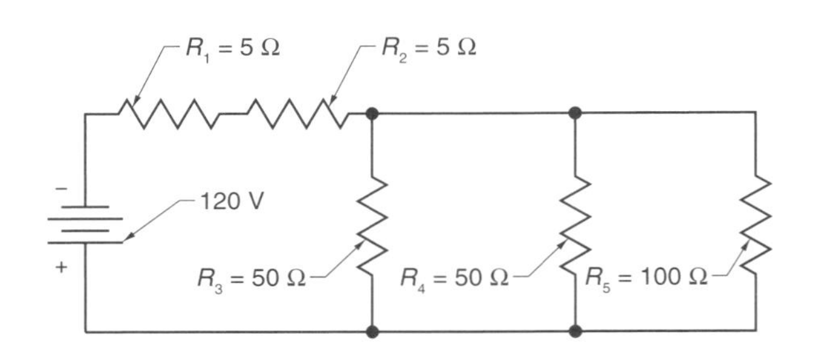

________ **11.** R_T = ___ Ω

________ **12.** I_1 = ___ A

________ **13.** V_D = ___ V

________ **14.** V_P = ___ V

________ **15.** P_{S1} = ___ W

________ **16.** P_{S2} = ___ W

________ **17.** P_{P1} = ___ W

________ **18.** P_{P2} = ___ W

________ **19.** P_{P3} = ___ W

________ **20.** P_T = ___ W

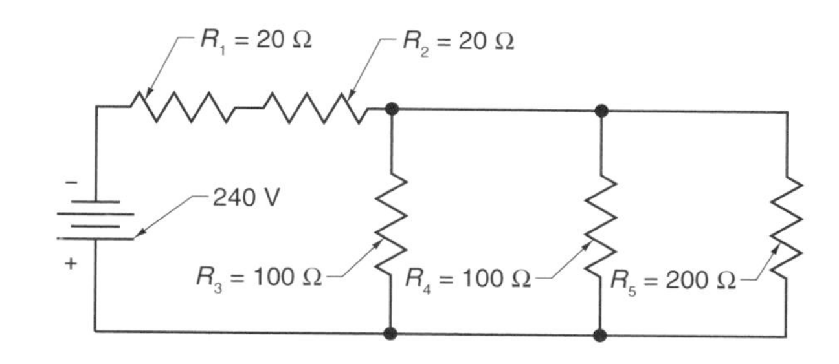

Identifying Connections

Identify each of the following as a series, parallel, or series/parallel connection.

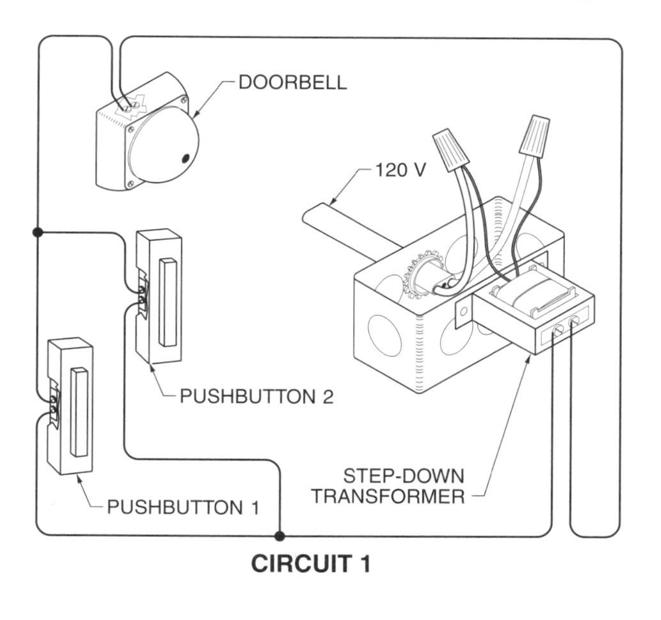

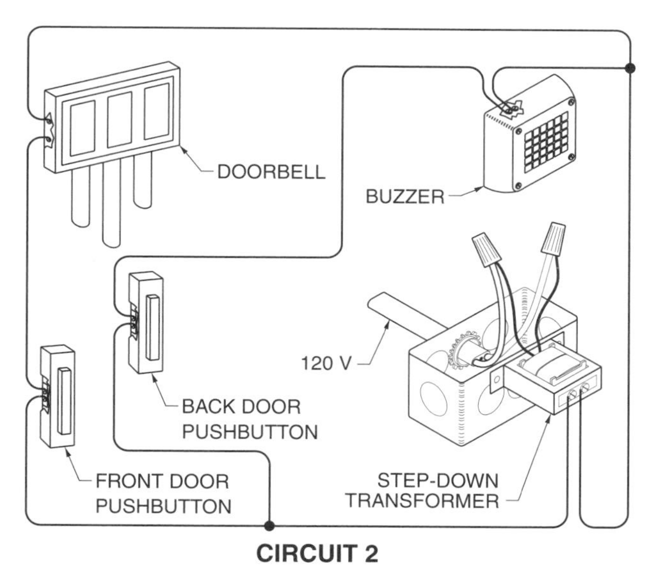

________ **1.** Pushbutton 1 and pushbutton 2 in Circuit 1 are connected in ___.

________ **2.** Pushbutton 1, pushbutton 2, and the doorbell in Circuit 1 are connected in ___.

________ **3.** The buzzer and the back door pushbutton in Circuit 2 are connected in ___.

________ **4.** The back door pushbutton, the doorbell, and the buzzer in Circuit 2 are connected in ___.

Simplifying a Series/Parallel Circuit

1. Combine the resistances and redraw the circuit below, step-by-step, until only one resistance (R_T) remains. State the value of total resistance. (Note: There should be five steps total.)

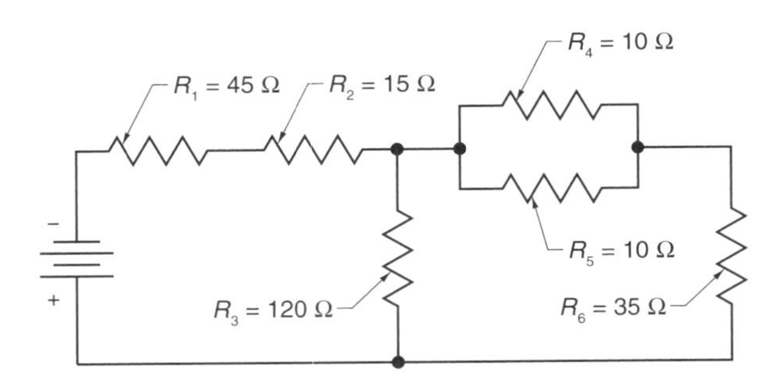

Power, Resistance, and Current in Series/Parallel Circuits

______________ **1.** P_{light} = ___ W

______________ **2.** P_{motor} = ___ W

______________ **3.** P_T = ___ W

______________ **4.** I_1 = ___ A

______________ **5.** I_2 = ___ A

______________ **6.** I_3 = ___ A

______________ **7.** I_T = ___ A

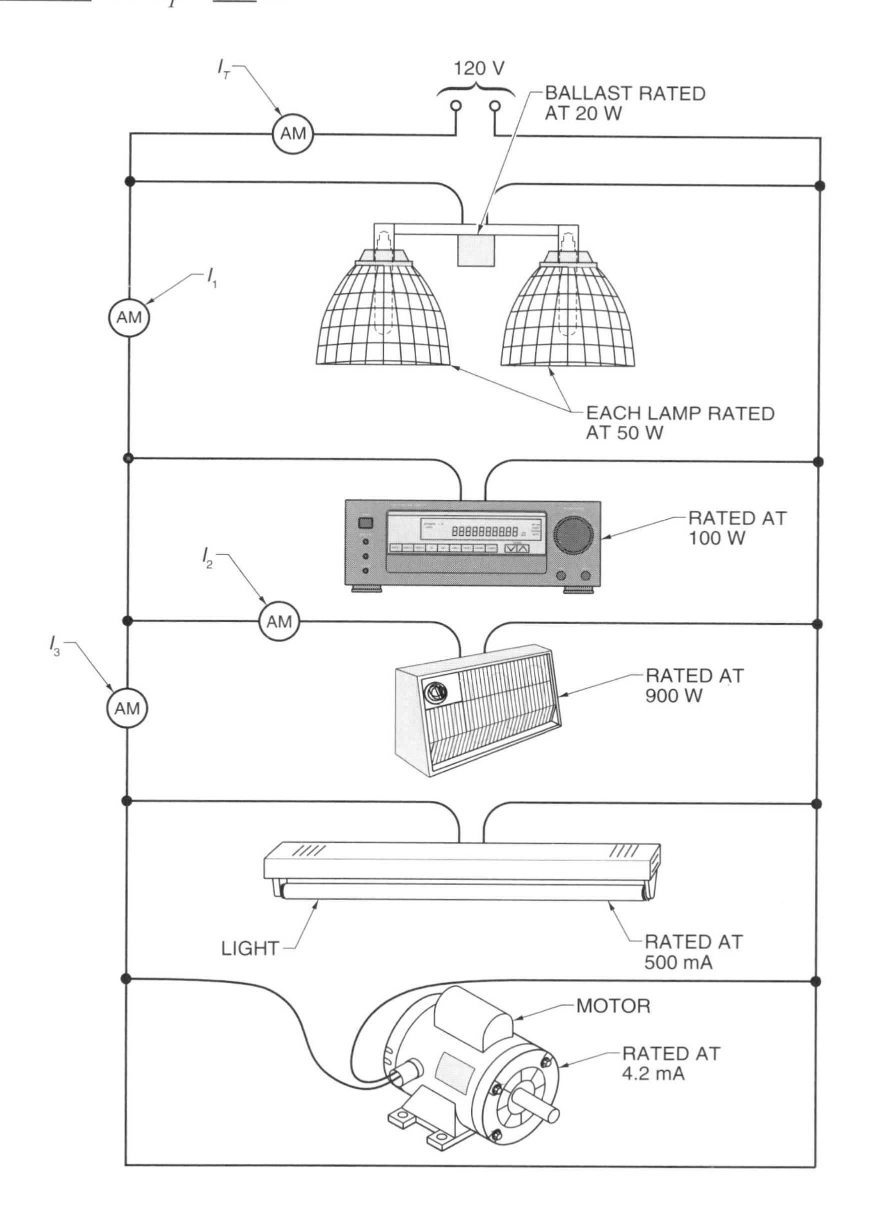

Complex Network Analysis Techniques

8

Name ______________________________ Date ______________

True-False

T F **1.** Current flow exiting a source voltage will return to the same source voltage.

T F **2.** Potential difference is the algebraic sum in potential between three different points in a circuit.

T F **3.** In Kirchhoff's voltage law, the sign of each voltage indicates if the voltage increases or decreases the loop voltage.

T F **4.** Two-voltage-source T-circuits have three voltage sources.

T F **5.** Bridge circuits are used only for controlling circuits.

T F **6.** Thevenin's theorem reduces a complex circuit to one voltage source with a series resistance.

T F **7.** Open-circuit voltage is always greater than when the voltage source is under load.

T F **8.** Kirchhoff's voltage law can be used any time there is voltage present in a closed-loop circuit.

T F **9.** In Thevenin's theorem, the load is considered part of the voltage source.

T F **10.** The ideal voltage source is used when making current calculations because in most cases, the values are acceptable.

Multiple Choice

______________ **1.** Kirchhoff's first and second laws are known as Kirchhoff's ___ law and Kirchhoff's voltage law.

A. resistance
B. current
C. impedance
D. none of the above

__________ **2.** A(n) ___ resistance bridge has resistances adjusted so there is equal current flow through the legs of the bridge and zero potential across the bridge.
A. balanced
B. unbalanced
C. complete
D. none of the above

__________ **3.** The ___ analyzes a circuit with multiple voltage sources by studying one voltage source at a time first and then combining the results.
A. Pythagorean theorem
B. Thevenin theorem
C. superposition theorem
D. Wheatstone bridge

__________ **4.** ___ theorem is a method of circuit analysis that reduces complex and basic circuits into a circuit with one current source and a parallel resistance.
A. Thevenin
B. Superposition
C. Ohm's
D. Norton's

__________ **5.** A Wheatstone bridge schematic diagram is usually drawn with four resistances connected in a ___-shaped configuration.
A. diamond
B. square
C. rectangle
D. none of the above

__________ **6.** Superposition theorem calculations can be performed using ___.
A. rules for series circuits
B. Kirchhoff's laws
C. Ohm's law
D. all of the above

Completion

__________ **1.** ___ is the voltage at a point in a circuit with respect to another point in the same circuit.

__________ **2.** A(n) ___ equation is a network equation based on Kirchhoff's current law.

__________ **3.** A(n) ___ bridge is a circuit used to take precise measurements of resistance.

__________ **4.** A(n) ___ resistance bridge has fixed resistances and the voltage across the bridge is proportional to the temperature of a variable resistor.

__________ **5.** Simplifying a complex circuit into an equivalent circuit containing one voltage source and resistance in series is known as ___.

Calculating Current in a Multiple-Voltage-Source Circuit

Calculate the resistor current for each voltage source and the total current for the following circuits:

A multiple-voltage-source circuit with voltage source values of 10 V and 6 V and resistance values of 3 Ω and 1 Ω.

__________ **1.** I_{RV1} = ___ A

__________ **2.** I_{RV2} = ___ A

__________ **3.** I_{RT} = ___ A

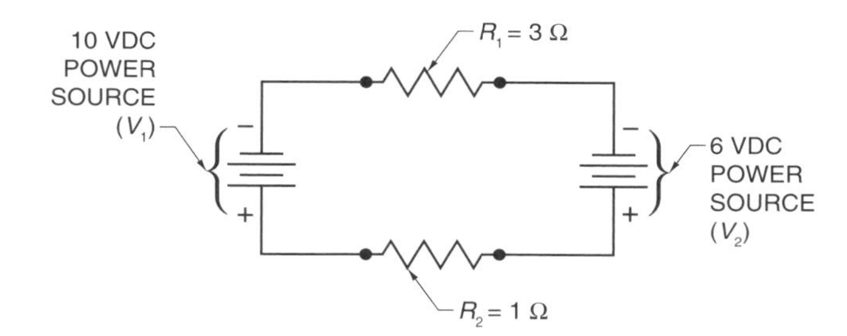

A multiple-voltage-source circuit with voltage source values of 18 V and 12 V and resistance values of 6 Ω and 5 Ω.

__________ **4.** I_{RV1} = ___ A

__________ **5.** I_{RV2} = ___ A

__________ **6.** I_{RT} = ___ A

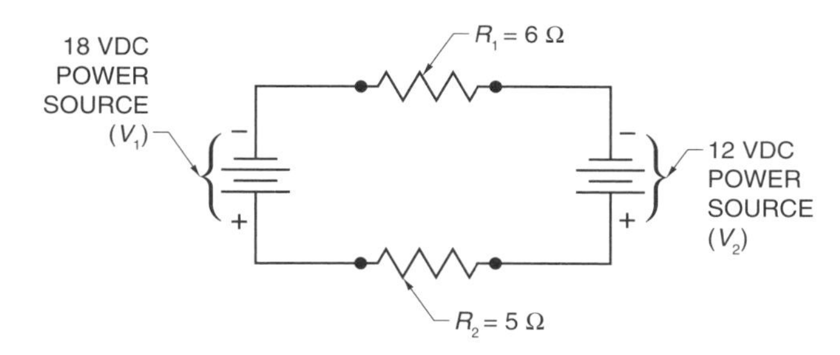

A multiple-voltage-source circuit with voltage source values of 12 V and 8 V and resistance values of 2 Ω and 3 Ω.

__________ **7.** I_{RV1} = ___ A

__________ **8.** I_{RV2} = ___ A

__________ **9.** I_{RT} = ___ A

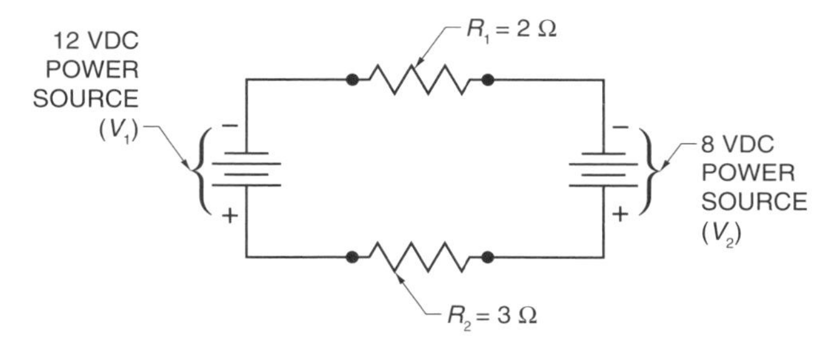

Thevenizing a Circuit

Calculate the individual voltages, Thevenin's resistance, load voltage, and load current for the following Wheatstone bridge circuits:

_______________ 1. V_{R2} = ___ V

_______________ 2. V_{R4} = ___ V

_______________ 3. V_{AB} = ___ V

_______________ 4. R_{TH} = ___ Ω

_______________ 5. V_L = ___ V

_______________ 6. I_L = ___ mA

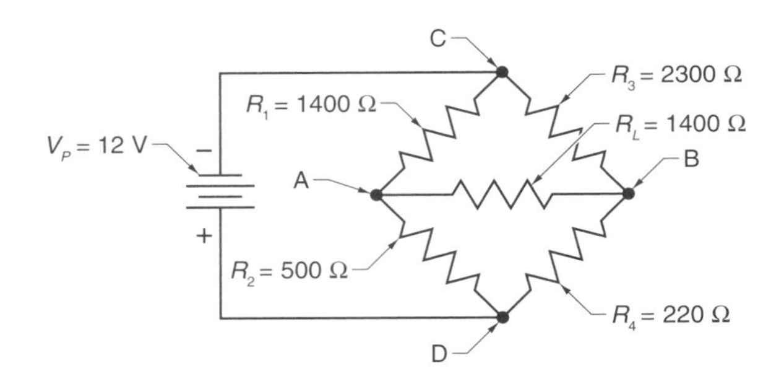

_______________ 7. V_{R2} = ___ V

_______________ 8. V_{R4} = ___ V

_______________ 9. V_{AB} = ___ V

_______________ 10. R_{TH} = ___ Ω

_______________ 11. V_L = ___ V

_______________ 12. I_L = ___ mA

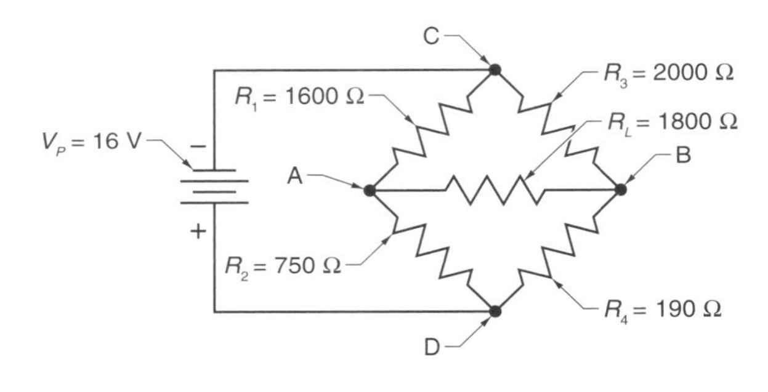

_______________ 13. V_{R2} = ___ V

_______________ 14. V_{R4} = ___ V

_______________ 15. V_{AB} = ___ V

_______________ 16. R_{TH} = ___ Ω

_______________ 17. V_L = ___ V

_______________ 18. I_L = ___ mA

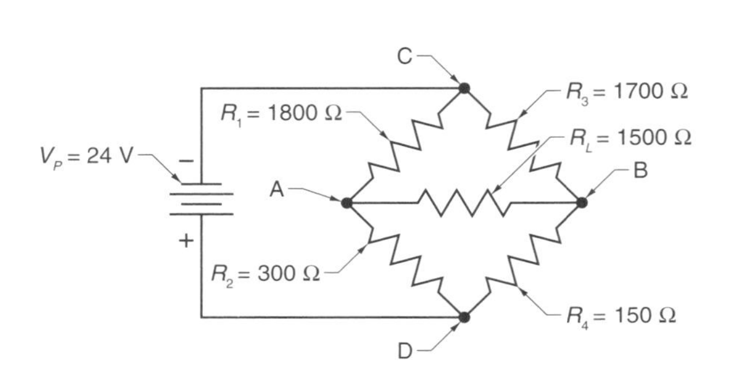

Wheatstone Bridges, I

1. Consider the Wheatstone bridge circuit pictured below. Thevenize the circuit into an equivalent circuit with a single power source and two resistors connected in series. Redraw the circuit step-by-step to show the process of thevenizing.

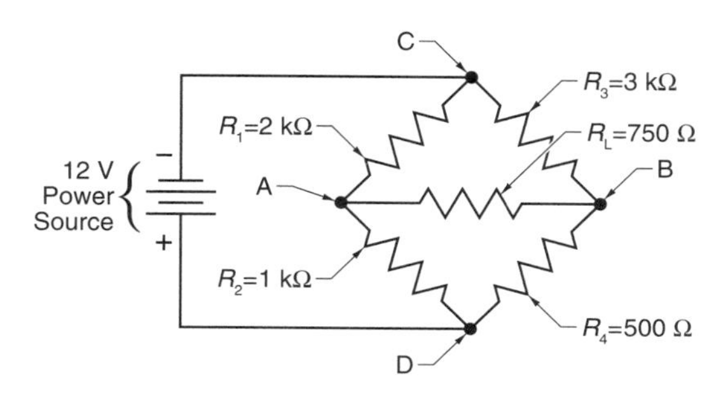

Solve for and state the values of V_{TH} and R_{TH}.

______________ **2.** V_{TH} = ___ V

______________ **3.** R_{TH} = ___ Ω

Wheatstone Bridges, II

Consider the bridge circuit below. An amplifier input module, which has an input impedance of 300 Ω, measures small changes in current.

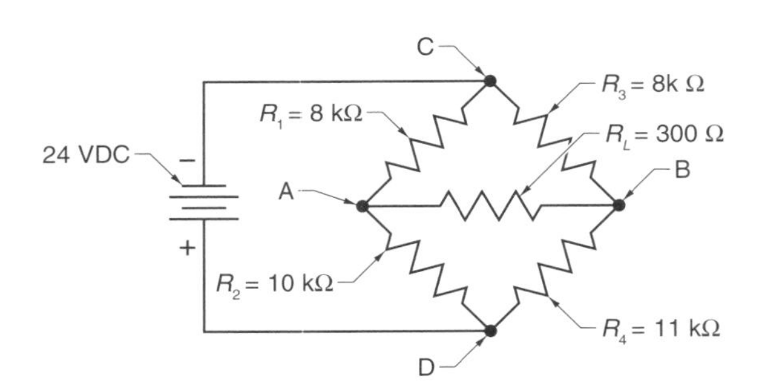

Solve for V_L, R_{TH}, and I_L in the circuit.

________________ **1.** V_L = ___ V

________________ **2.** R_{TH} = ___ Ω

________________ **3.** I_L = ___ mA

Electromagnetism 9

Name ______________________________ Date ______________

True-False

T F **1.** An inductor is any material through which current flows easily.

T F **2.** A coil is a circular wound wire of insulated conductors that produces lines of magnetic flux.

T F **3.** The polarity of a coil is determined by applying the left-hand rule for coils.

T F **4.** In terms of electromagnets, air provides a better path for magnetic lines than soft iron.

T F **5.** Transformers depend on magnetism to operate correctly.

T F **6.** Rowland's law is similar to Ohm's law and is used for electrical circuits.

T F **7.** Magnetomotive force in a magnetic circuit is similar to voltage in an electrical circuit.

T F **8.** Reluctance is the opposition to the flow of magnetic lines of force.

T F **9.** A gauss is equal to one line of magnetism per square meter.

T F **10.** Magnetic saturation occurs when an increase in voltage will not cause an increase in the number of magnetic lines of force.

T F **11.** Retentivity is highest in soft iron and lowest in hard steel.

T F **12.** Decreasing resistance in a conductor yields higher current flow and more magnetic lines of force.

T F **13.** Lenz's law determines the direction of an induced magnetomotive force.

T F **14.** The angle of cutting motion when a conductor cuts the magnetic field affects the amount of induced voltage.

T F **15.** Magnetic lines of force always move from a weak magnetic field toward a strong one.

Multiple Choice

________ **1.** Electromagnetism is the magnetism produced when electricity passes through a(n) ___.
- A. meter
- B. resistor
- C. inductor
- D. conductor

________ **2.** The ___ rule for conductors can be used to determine the direction of magnetic lines of force.
- A. right-hand
- B. left-hand
- C. one-hand
- D. none of the above

________ **3.** Magnetic circuit paths may be ___.
- A. series/parallel
- B. series
- C. parallel
- D. all of the above

________ **4.** Rowland's law involves all of the following variables except ___.
- A. magnetomotive force
- B. reluctance
- C. inductance
- D. magnetic lines of force

________ **5.** A(n) ___ is the unit of total magnetic lines of force.
- A. weber
- B. maxwell
- C. ohm
- D. joule

________ **6.** A material's ability to carry magnetic lines of force is referred to as ___.
- A. flux
- B. permeability
- C. electromagnetism
- D. reluctance

________ **7.** ___ motion is the speed at which a conductor cuts a magnetic field.
- A. Derivative
- B. Angle
- C. Relative
- D. none of the above

_______________ **8.** Field windings are magnets used to produce the magnetic field in a ___.
   A. motor
   B. transformer
   C. generator
   D. motion sensor

_______________ **9.** The strength of a coil's magnetic field is affected by all of the following factors except ___.
   A. number of conductor turns
   B. amount of current flow
   C. type of core material
   D. hysteresis loss

_______________ **10.** ___ is the property of a core material that allows the passage of magnetic lines of force.
   A. Permeance
   B. Reluctivity
   C. Permeability
   D. Inductance

Completion

_______________ **1.** Magnetic lines of force traveling in the same direction ___ each other.

_______________ **2.** Electromagnets are coils that have a core material made of soft ___.

_______________ **3.** A(n) ___ circuit is the path taken by the magnetic lines of force.

_______________ **4.** A(n) ___ is the unit of flux defined as one line of flux.

_______________ **5.** Total reluctance in a circuit is calculated by ___ all circuit reluctances.

_______________ **6.** ___ is the opposition a circuit has to a change in current due to energy stored in a magnetic field.

_______________ **7.** ___ current is unwanted current caused by the rate of change in the induced magnetic field.

_______________ **8.** A ripple is a change in DC output ___ level.

_______________ **9.** The magnetizing ___ intensity measures the magnetic strength per unit length.

_______________ **10.** Hysteresis is the ___ of magnetic lines of force behind the magnetizing force that causes them.

Intensity of Magnetizing Force

Calculate the magnetomotive force (mmf) and the intensity of the magnetizing force.

A conductor carrying 200 mA with 1000 turns over a 5 cm length.

_______________ **1.** F_m = ___ Gb

_______________ **2.** H = ___ Oe

A conductor carrying 100 mA with 500 turns over a 5 cm length.

_______________ **3.** F_m = ___ Gb

_______________ **4.** H = ___ Oe

A conductor carrying 300 mA with 250 turns over a 15 cm length.

_______________ **5.** F_m = ___ Gb

_______________ **6.** H = ___ Oe

A conductor carrying 500 mA with 450 turns over a 25 cm length.

_______________ **7.** F_m = ___ Gb

_______________ **8.** H = ___ Oe

Solenoids

A solenoid is an electric output device that converts electrical energy into a linear mechanical force. The current drawn by a solenoid coil depends on the applied voltage and size of the coil. Manufacturers list coil specifications to assist in installation and sizing of components. Because magnetic coils are encapsulated and cannot be repaired, they must be replaced when they fail.

COIL SPECIFICATIONS

Size	Number of poles	Inrush current* 60 cycles					Sealed current* 60 cycles					Approximate operating time†	
		120 V	208 V	240 V	480 V	600 V	120 V	208 V	240 V	480 V	600 V	Pick-up	Drop-out
00	1-2-3	0.50	0.29	0.25	0.12	0.07	0.12	0.07	0.06	0.03	0.02	28	13
0	1-2-3-4	0.88	0.50	0.44	0.22	0.17	0.14	0.08	0.07	0.04	0.03	29	14
1	1-2-3-4	1.54	0.89	0.77	0.39	0.31	0.18	0.10	0.09	0.04	0.04	26	17
2	2-3-4	1.80	1.04	0.90	0.45	0.36	0.25	0.14	0.13	0.06	0.05	32	14
3	2-3	4.82	2.78	2.41	1.21	0.97	0.36	0.21	0.18	0.09	0.07	35	18
	4	5.34	3.08	2.67	1.33	1.07	0.39	0.23	0.20	0.10	0.08	35	18
4	2-3	8.30	4.80	4.15	2.08	1.66	0.54	0.31	0.27	0.14	0.11	41	18
	4	9.90	5.71	4.95	2.47	1.98	0.61	0.35	0.31	0.15	0.12	41	18
5	2-3	16.23	9.36	8.11	4.06	3.25	0.81	0.47	0.41	0.20	0.16	43	18

* in A

† in ms

For example, the coil used in a size 2 motor starter may have two, three, or four poles and draw 0.13 A sealed current when connected to a 240 V power source. This current value is used when selecting circuit fuse, wire, and transformer sizes.

Using the Coil Specifications provided, answer the following questions.

A 120 V, size 00, three-pole contactor

__________ **1.** The inrush current is ___ A.

__________ **2.** The sealed current is ___ A.

__________ **3.** The pick-up time is ___ ms.

__________ **4.** The drop-out time is ___ ms.

A 240 V, size 5, three-pole contactor

__________ **5.** The inrush current is ___ A.

__________ **6.** The sealed current is ___ A.

__________ **7.** The pick-up time is ___ ms.

__________ **8.** The drop-out time is ___ ms.

A 120 V, size 3, three-pole contactor

__________ **9.** The inrush current is ___ A.

__________ **10.** The sealed current is ___ A.

__________ **11.** The pick-up time is ___ ms.

__________ **12.** The drop-out time is ___ ms.

A 480 V, size 3, three-pole contactor

__________ **13.** The inrush current is ___ A.

__________ **14.** The sealed current is ___ A.

__________ **15.** The pick-up time is ___ ms.

__________ **16.** The drop-out time is ___ ms.

DC Circuit Inductance 10

True-False

T F 1. Electromotive force is defined as electrical pressure applied to a circuit.

T F 2. A greater number of coil layers on a core will produce a lower inductance.

T F 3. A coiled conductor has more inductance than a straight conductor.

T F 4. Wheeler's formula is the formula used to determine inductance for each inductor coil style.

T F 5. Multilayer coils are referred to as solenoids.

T F 6. Unlike other coils, the magnetic field of a toroid coil is confined to its core.

T F 7. Radio frequency chokes are designed for low impedance over small frequency ranges.

T F 8. An inductor may be shielded in an electrical circuit to prevent the interaction of magnetic fields.

T F 9. Noninductive coil inductors can be used in an application that requires a circuit to be resistive in nature.

T F 10. Inductors are usually rated in terms of inductance measured in ohms.

T F 11. A resistive circuit contains only inductance.

T F 12. Mutual inductance is measured in henrys.

T F 13. Inductors are only connected in series.

T F 14. A pancake coil is also known as a flat spiral coil.

Multiple Choice

______ **1.** ___ force always opposes current change in a conductor.
A. Electromotive
B. Counter-electromotive
C. Magnetic
D. none of the above

______ **2.** The ___ is the unit of inductance equal to the voltage induced per rate of current change.
A. joule
B. gilbert
C. maxwell
D. henry

______ **3.** Inductance is affected by all of the following except core ___.
A. diameter
B. material
C. shape
D. none of the above

______ **4.** A ___ coil is a short coil with only a few conductor turns along its length.
A. pancake
B. single
C. multilayer
D. wound

______ **5.** A ___ inductor is usually covered in plastic with a conductor at each end for connecting or soldering into a circuit.
A. variable
B. molded
C. shielded
D. noninductive coil

______ **6.** A(n) ___ inductive circuit has an inductor connected to a battery through a switch.
A. real
B. model
C. ideal
D. none of the above

______ **7.** ___ inductance is the effect of one coil inducing a voltage into another coil.
A. Common
B. Actual
C. Mutual
D. Joined

Completion

____________________ **1.** ___ is the property of a conductor to induce voltage within itself due to changes in current.

____________________ **2.** The ability of a material to carry magnetic lines of force is referred to as ___.

____________________ **3.** A(n) ___ coil has more than one layer of turns and is wound on an air-core form.

____________________ **4.** A(n) ___ inductor has an inductance varied by its core.

____________________ **5.** Inductive ___ is the opposition of an inductor to AC.

____________________ **6.** A(n) ___ inductive circuit has both resistance and inductance.

____________________ **7.** High induced voltage is an energy release caused by the ___ in an inductive circuit.

____________________ **8.** When inductors are connected in parallel, total circuit current ___.

Identification—Coils

____________________ **1.** Pancake

____________________ **2.** Single-layer

____________________ **3.** Multilayer

____________________ **4.** Toroid

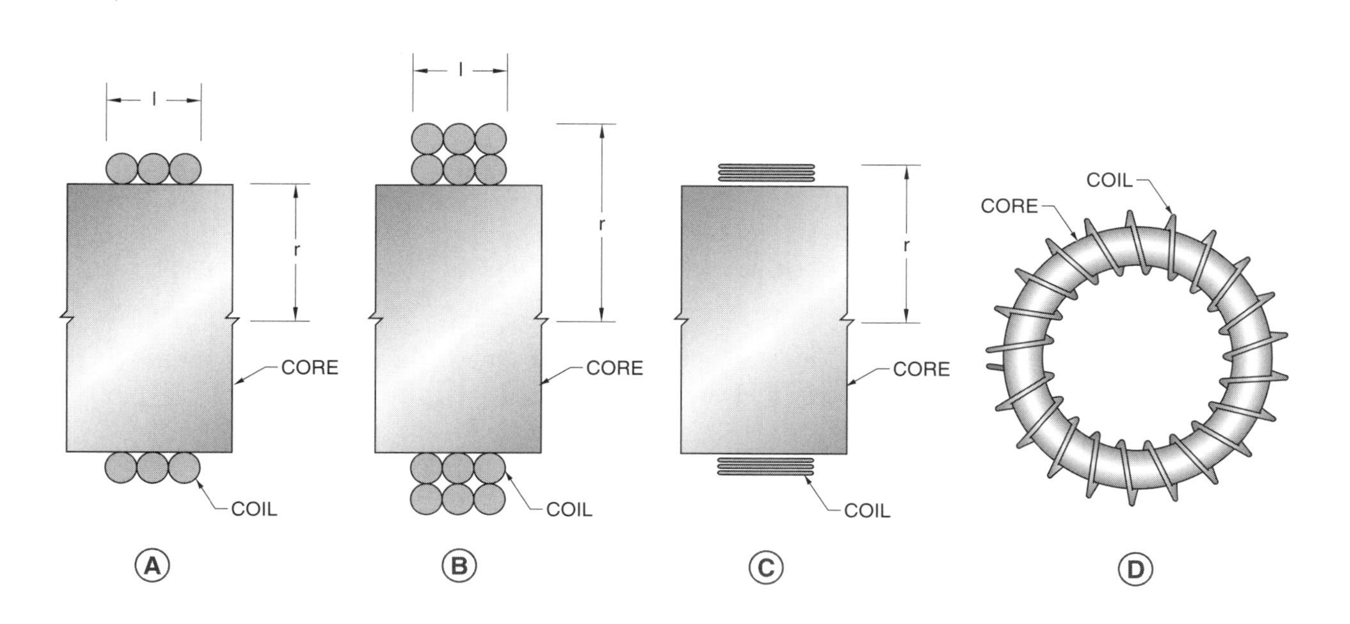

Calculating Inductance

Calculate the value of inductance for the following:

________________ **1.** A single-layer coil with a radius of 0.5″, a length of 6″, and 12 turns

________________ **2.** A single-layer coil with a radius of 0.7″, a length of 10″, and 10 turns

________________ **3.** A single-layer coil with a radius of 1″, a length of 4″, and 12 turns

________________ **4.** A single-layer coil with a radius of 1″, a length of 4″, and 8 turns

________________ **5.** A multilayer coil with 40 turns on a radius air cylinder of 0.2″, a length of 2″, and a coil depth of 0.5″

________________ **6.** A multilayer coil with 60 turns on a radius air cylinder of 0.1″, a length of 3″, and a coil depth of 0.8″

________________ **7.** A multilayer coil with 50 turns on a radius air cylinder of 0.1″, a length of 2″, and a coil depth of 0.4″

________________ **8.** A multilayer coil with 30 turns on a radius air cylinder of 0.1″, a length of 1″, and a coil depth of 0.5″

________________ **9.** A pancake coil with 20 turns, a radius of 0.1″, and a coil depth of 0.1″

________________ **10.** A pancake coil with 35 turns, a radius of 0.2″, and a coil depth of 0.2″

________________ **11.** A pancake coil with 15 turns, a radius of 0.3″, and a coil depth of 0.2″

________________ **12.** A pancake coil with 30 turns, a radius of 0.125″, and a coil depth of 0.1″

Calculating Mutual Inductance

Calculate the value of mutual inductance for the following:

________________ **1.** Two coils, each with an inductance of 3 H, with a coefficient of coupling of 50%

________________ **2.** Two coils, each with an inductance of 4 H, with a coefficient of coupling of 75%

________________ **3.** Two coils, the first with an inductance of 2 H and the second with an inductance of 5 H, with a coefficient coupling of 50%

________________ **4.** Two coils, the first with an inductance of 3 H and the second with an inductance of 4 H, with a coefficient coupling of 75%

Calculating Total Inductance—Series Connections

Calculate the value of total inductance for the following:

________________ **1.** Two inductors connected in series, each with an inductance of 2 H, and aiding magnetic fields with a coefficient coupling of 75%

________________ **2.** Two inductors connected in series, the first with an inductance of 1 H and the second with an inductance of 3 H, and aiding magnetic fields with a coefficient coupling of 50%

________________ **3.** Two inductors connected in series, the first with an inductance of 2 H and the second with an inductance of 3 H, and aiding magnetic fields with a coefficient coupling of 75%

________________ **4.** Two inductors connected in series, the first with an inductance of 2 H and the second with an inductance of 4 H, and aiding magnetic fields with a coefficient coupling of 50%

________________ **5.** Two inductors connected in series, each with inductance of 2 H, and opposing magnetic fields with a coefficient coupling of 75%

________________ **6.** Two inductors connected in series, the first with an inductance of 1 H and the second with an inductance of 3 H, and opposing magnetic fields with a coefficient coupling of 50%

________________ **7.** Two inductors connected in series, the first with an inductance of 2 H and the second with an inductance of 3 H, and opposing magnetic fields with a coefficient coupling of 75%

________________ **8.** Two inductors connected in series, the first with an inductance of 2 H and the second with an inductance of 4 H, and opposing magnetic fields with a coefficient coupling of 50%

Calculating Total Inductance—Parallel Connections

Calculate the value of total inductance for the following:

________________ **1.** Two inductors connected in parallel, with inductance values of 0.2 H and 0.4 H, with aiding mutual induction of 0.13 H and a coefficient coupling of 50%

________________ **2.** Two inductors connected in parallel, with inductance values of 0.3 H and 0.6 H, with aiding mutual induction of 0.13 H and a coefficient coupling of 50%

________________ **3.** Two inductors connected in parallel, with inductance values of 0.4 H and 0.5 H, with aiding mutual induction of 0.14 H and a coefficient coupling of 50%

______________ **4.** Two inductors connected in parallel, with inductance values of 0.1 H and 0.4 H, with aiding mutual induction of 0.14 H and a coefficient coupling of 50%

______________ **5.** Two inductors connected in parallel, with inductance values of 2 H and 4 H, with an opposing mutual induction of 0.35 H and a coefficient coupling of 15%

______________ **6.** Two inductors connected in parallel, with inductance values of 1 H and 4 H, with an opposing mutual induction of 0.45 H and a coefficient coupling of 15%

______________ **7.** Two inductors connected in parallel, with inductance values of 2 H and 3 H, with an opposing mutual induction of 0.45 H and a coefficient coupling of 15%

______________ **8.** Two inductors connected in parallel, with inductance values of 3 H and 5 H, with an opposing mutual induction of 0.40 H and a coefficient coupling of 15%

Mutual Inductance

Consider two coils of a transformer. The coils are spaced far apart. As a result, the coils have a low coefficient of coupling. The inductances of the coils are 4 H and 5 H, respectively. The coefficient of coupling is 60%.

What is the mutual inductance of the coils?

______________ **1.** M = ___ H

What is the total inductance of the coils?

______________ **2.** L_T = ___ H

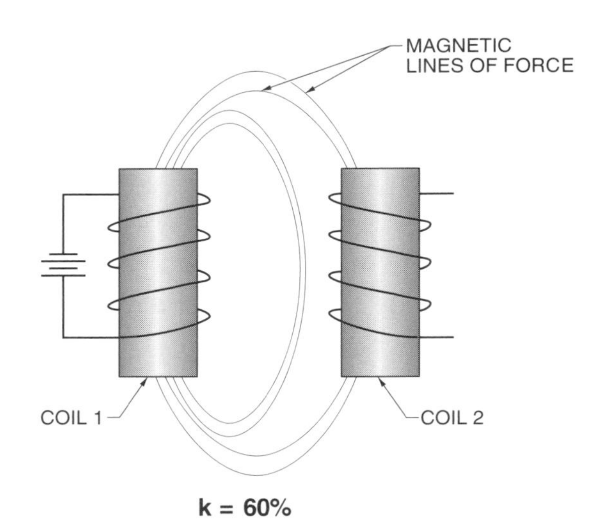

DC Circuit Capacitance 11

Name ____________________ Date ____________

True-False

T F **1.** Capacitance is the ability of a component or circuit to store energy in the form of an electrical charge.

T F **2.** In a capacitive circuit with DC voltage applied, current flows when capacitive voltage equals the source voltage.

T F **3.** Because the farad is too large a unit of measure for average capacitor applications, picofarads and microfarads are often used.

T F **4.** Capacitance is directly proportional to the size of the capacitor plates.

T F **5.** There are three types of capacitors.

T F **6.** Paper capacitors are waxed in a paraffin coating for electrical insulation.

T F **7.** Ceramic capacitors are often referred to as self-healing capacitors.

T F **8.** Compared to other types of capacitors, aluminum electrolytic capacitors have low leakage current.

T F **9.** A supercapacitor is often used as a battery replacement because it has characteristics similar to those of a battery.

T F **10.** Tubular paper capacitors have six colored bands used for identification purposes.

T F **11.** Capacitors can be connected in a series or parallel circuit but not in a series/parallel combination circuit.

T F **12.** Dielectric constant and dielectric strength are the same thing.

T F **13.** A polarized capacitor has separate negative and positive connection leads.

T F **14.** Displacement current is current that exists in addition to normal current.

T F **15.** The thickness of the dielectric of a capacitor does not affect total capacitance.

T F **16.** Capacitivity is another name for dielectric breakdown.

T F **17.** Insulation can be classified as solid, fluid, or gas.

T F **18.** Leaded capacitors are a form of aluminum electrolytic capacitors.

Multiple Choice

______________ **1.** A ___ is the quantity of electric charge that passes any cross-section of a conductor in one second when the current is maintained at 1 A.
- A. farad
- B. gilbert
- C. coulomb
- D. maxwell

______________ **2.** Capacitance of a circuit is affected by all of the following except ___.
- A. dielectric constant of the material
- B. surface area of the plates
- C. thickness of the material
- D. shape of the plates

______________ **3.** Dielectric ___ is the ratio of the capacitance of a capacitor with a given dielectric to the capacitance of a capacitor with air as the dielectric.
- A. strength
- B. constant
- C. breakdown
- D. resistance

______________ **4.** The most commonly used capacitors are ___ capacitors.
- A. ceramic
- B. mica
- C. paper
- D. oil

______________ **5.** A(n) ___ is an electrically charged atom.
- A. electrolyte
- B. neutron
- C. ion
- D. nucleus

________ **6.** A capacitor that offers high capacitance in a small volume is a ___.
- A. trimmer capacitor
- B. supercapacitor
- C. paper capacitor
- D. padder capacitor

________ **7.** Dielectric leakage current results from the dielectric having high ___.
- A. resistance
- B. impedance
- C. capacitance
- D. inductance

________ **8.** The two types of capacitor losses are dielectric leakage current and ___ loss.
- A. energy
- B. voltage
- C. power
- D. hysteresis

________ **9.** A(n) ___ can be used to test the reliability of a capacitor.
- A. ammeter
- B. voltmeter
- C. ohmmeter
- D. none of the above

________ **10.** As capacitor plate size increases, capacitance value ___.
- A. increases
- B. decreases
- C. stays the same
- D. none of the above

Completion

________ **1.** ___ current is the small amount of current that flows through insulation.

________ **2.** ___ breakdown is the breakdown of insulation between the plates of a capacitor.

________ **3.** Most paper capacitors are made in a(n) ___ design.

________ **4.** ___ capacitors are sometimes referred to as glass capacitors.

________ **5.** A(n) ___ capacitor is constructed so that the capacitance values within the capacitor can be changed.

________________ **6.** A resistive-capacitive ___ is the amount of time required for the capacitor to charge to 63.2% of the maximum voltage across the circuit.

________________ **7.** For circuits with capacitors connected in series/parallel, values of the ___-connected capacitors are determined first.

________________ **8.** Capacitor losses typically occur through the ___.

________________ **9.** Capacitance of parallel plates is dependent on the type of ___ material used.

________________ **10.** A(n) ___ capacitor is a small adjustable capacitor that is connected in series in the tuning circuit of a radio.

Calculating Capacitance

Calculate the value of capacitance for the following:

________________ **1.** A resin capacitor with a dielectric constant of 2.5, with 3″ by 3″ plates separated by 0.010″

________________ **2.** A glass capacitor with a dielectric constant of 5.0, with 2″ by 2″ plates separated by 0.020″

________________ **3.** A ceramic capacitor with a dielectric constant of 5.7, with 3″ by 2″ plates separated by 0.010″

________________ **4.** A rubber capacitor with a dielectric constant of 4.6, with 4″ by 2″ plates separated by 0.020″

Calculating Maximum Capacitance

Calculate the maximum capacitance for the following:

________________ **1.** An 8-plate variable capacitor with each plate having an area of 0.5 in^2 and a distance of 0.2″ between the plates

________________ **2.** A 9-plate variable capacitor with each plate having an area of 0.4 in^2 and a distance of 0.1″ between the plates

________________ **3.** A 6-plate variable capacitor with each plate having an area of 0.3 in^2 and a distance of 0.1″ between the plates

________________ **4.** A 5-plate variable capacitor with each plate having an area of 0.5 in^2 and a distance of 0.2″ between the plates

Capacitors in Series

Calculate the total circuit capacitance for each circuit.

_______________ **1.** C_T = ___ μF

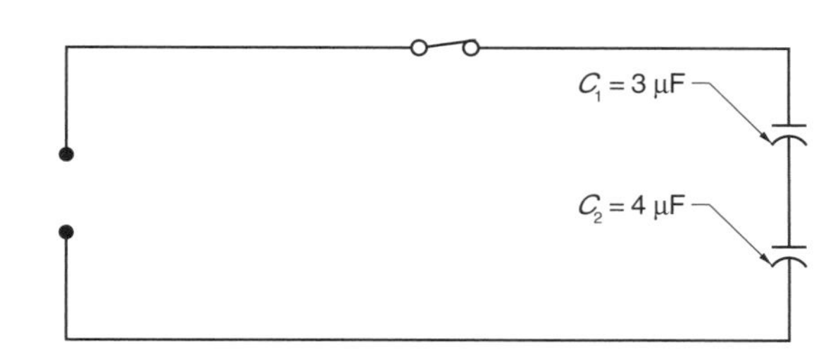

_______________ **2.** C_T = ___ μF

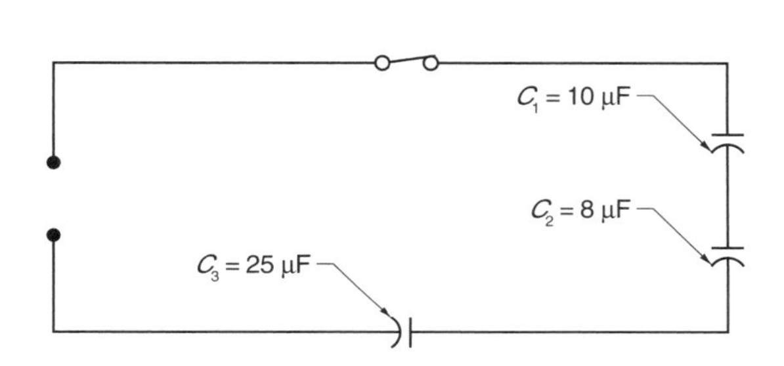

_______________ **3.** C_T = ___ μF

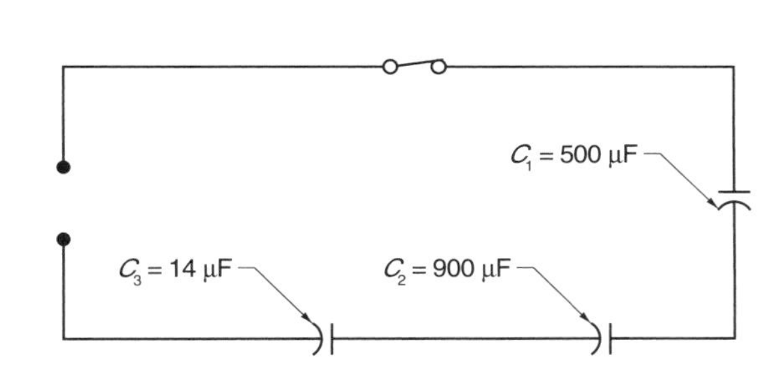

_______________ **4.** C_T = ___ μF

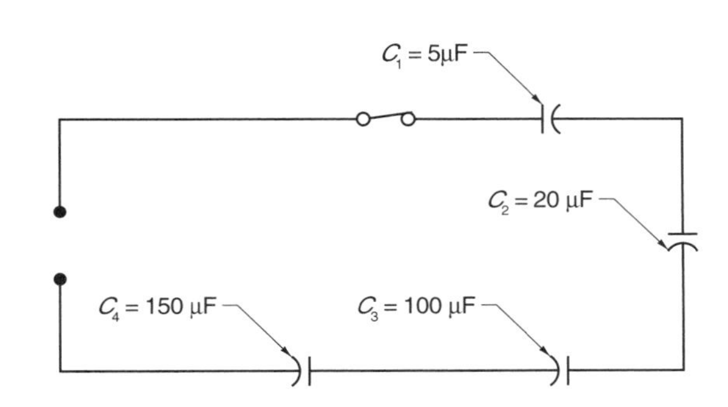

Capacitors in Parallel

Calculate the total circuit capacitance for each circuit.

________________ **1.** C_T = ___ μF

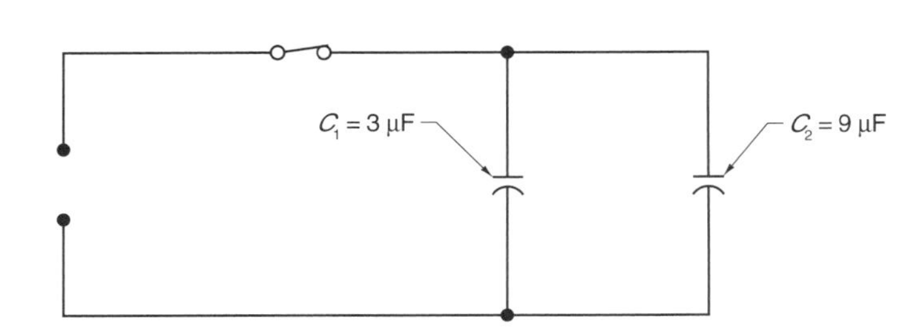

________________ **2.** C_T = ___ μF

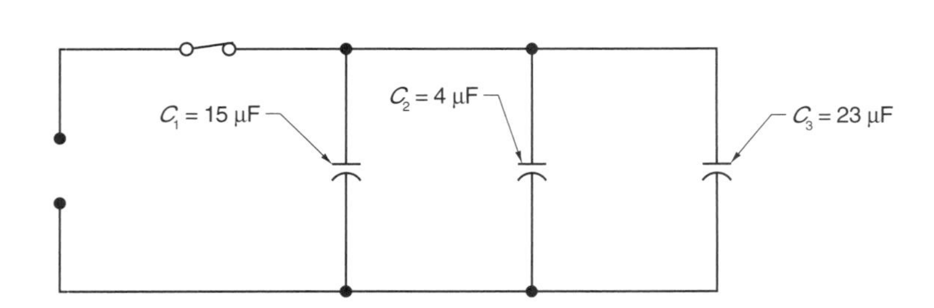

________________ **3.** C_T = ___ μF

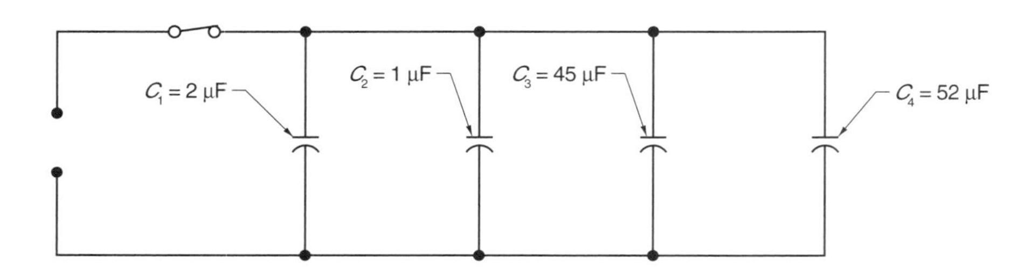

________________ **4.** C_T = ___ μF

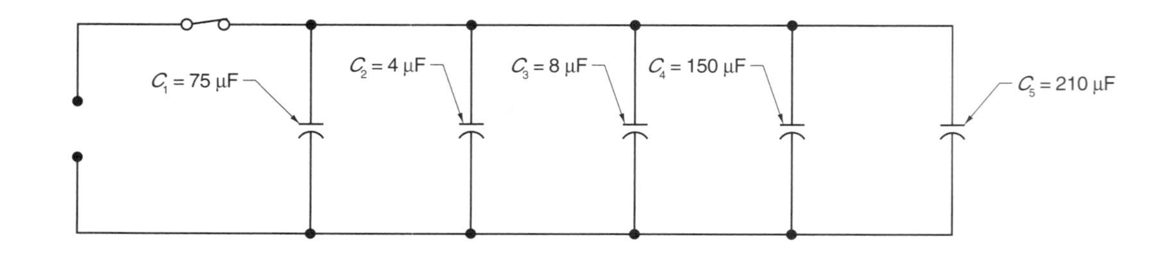

AC Fundamentals 12

Name ______________________________ Date ______________

True-False

T F **1.** DC flow occasionally reverses in direction, unlike AC.

T F **2.** An alternation is a complete cycle.

T F **3.** The wavelength of a signal and its frequency are proportional.

T F **4.** Skin effect occurs in AC when more current flows near the outer surface of a conductor at higher frequencies.

T F **5.** Phase is the time relationship of a sine wave to a known time period.

T F **6.** The part of the generator from which electric power is generated is known as the stator.

T F **7.** Steam-powered turbines can be used to drive electric power generators.

T F **8.** In AC circuits, voltage and current are always in phase.

T F **9.** Ignitrons are useful for their ability to withstand high voltages and conduct high currents.

T F **10.** Harmonics can either be voltage harmonics or current harmonics.

T F **11.** AC waveforms cross the 0 A baseline when there is a reversal of current flow.

T F **12.** A utilization voltage is a primary voltage of generators given to all loads.

T F **13.** Forward breakover voltage is the voltage needed to switch from a conductive state to a nonconductive state in a silicon-controlled rectifier.

T F **14.** Semiconductor devices make it easy to switch from AC to DC.

Multiple Choice

______________ **1.** A ___ is the time required to produce one complete cycle of a waveform.
A. cycle
B. period
C. rotation
D. moment

______________ **2.** The ___ is the international unit of frequency equal to 1 cycle per second.
A. hertz
B. gilbert
C. decibel
D. none of the above

______________ **3.** Alternators are rated in ___.
A. megawatts
B. joules
C. volt-amperes
D. ohms

______________ **4.** ___ are electrical devices that use electromagnetism to change voltage from one level to another.
A. Generators
B. Transformers
C. Motors
D. all of the above

______________ **5.** A(n) ___ is an electronic device that allows current to flow in only one direction.
A. cathode
B. anode
C. thyristor
D. diode

______________ **6.** ___ current is current that flows in only one direction.
A. Alternating
B. Harmonic
C. Direct
D. none of the above

______________ **7.** A ___ is the distance covered by one complete cycle of a given frequency of sound as it passes through the air.
A. wavelength
B. harmonic
C. cosine wave
D. phase

______________ **8.** DC waveforms can take all the shapes below except ___ waves.
A. sine
B. sawtooth
C. current
D. none of the above

Completion

______________ **1.** ___ power is more easily distributed at higher voltages and lower current.

______________ **2.** A(n) ___ is one complete positive and negative alternation of a waveform.

______________ **3.** ___ is the number of cycles per second in an AC sine wave.

______________ **4.** Wavelength is measured in ___.

______________ **5.** ___ is the process of changing AC to DC.

______________ **6.** An even ___ is an even multiple of a fundamental frequency.

______________ **7.** A(n) ___ conductor is composed of several strands of solid wire wrapped together to make a single conductor.

______________ **8.** Another name for an alternator is an AC ___.

______________ **9.** Ignitrons can be wired in ___ if greater current capacity is needed.

______________ **10.** A silicon-controlled rectifier is a solid-state rectifier that is capable of rapidly switching ___.

Identification—Frequency and Wavelength

Identify the frequency for each device, then calculate the wavelength.

____________________ **1.** 1.2 MHz, λ = ___ m

____________________ **2.** 220 MHz, λ = ___ m

____________________ **3.** 850 MHz, λ = ___ m

____________________ **4.** 5 GHz, λ = ___ m

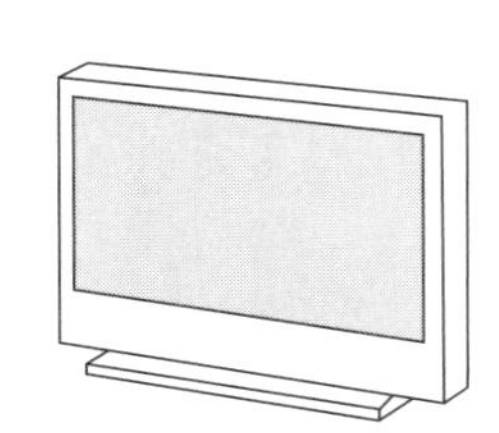

TELEVISION BROADCAST
(B)

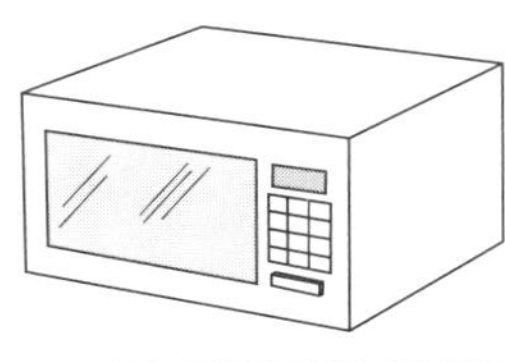

MICROWAVE OVEN
(C)

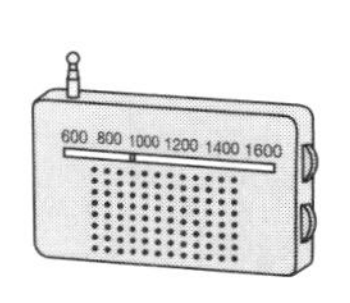

AM RADIO
(D)

Calculating Wavelength

Calculate the wavelength of the following:

An AM broadcast that has a frequency of 2 MHz

____________________ **1.** λ = ___ m

A TV broadcast that has a frequency of 150 MHz

____________________ **2.** λ = ___ m

A mobile phone that has a frequency of 1.6 GHz

____________________ **3.** λ = ___ m

A transmitter that has a period of 0.33 sec

____________________ **4.** λ = ___ m

A microwave that has a period of 0.5 sec

____________________ **5.** λ = ___ m

A beacon that has a period of 0.25 sec

____________________ **6.** λ = ___ m

Activating Frequency

Activating frequency is the limit of the number of pulses per second that can be detected by a photoelectric control in a given time period. All photoelectric controls have an activating frequency. To determine the required activating frequency of a photoelectric application, apply the following procedure:

1. Determine the maximum speed of the objects to be measured and convert that speed to the number of seconds it takes the object to travel 1″. For example, 21′ per min = 0.238 sec/in.
2. Calculate the dark input signal duration. This is the time period when the photosensor is dark because the detected object is blocking the light beam. Dark input signal duration is found by multiplying the minimum dimension of the object to be detected (in in.) by the speed in sec/in. For example, if a 6″ × 3″ object with the 3″ dimension in the line of travel has a 0.263 sec/in. value, the dark input signal duration is 0.789 sec (3 × 0.263).
3. Calculate the light input signal duration. This is the time period when the photosensor is lit because no detectable object is in the light beam. Light input signal duration is found by multi plying the minimum distance between the objects to be detected by the speed in sec/in. For example, if the moving objects are spaced a minimum of 4.00″ apart on a conveyor traveling at 0.200 sec/in., the light input signal duration is 0.800 sec (4.00 × 0.200).
4. Calculate the activating frequency. This is the sum of the dark input and the light input signal durations. For example, if the dark input signal duration is 0.789 sec and the light input signal duration is 0.800 sec, the activating frequency is 1.589 sec (0.789 sec + 0.800 sec).

Use the procedure above to answer the following questions:

A photoelectric sensor detects 2″ × 2″ objects that travel at 60 ft/min and are 5″ apart.

__________________ **1.** Speed of the objects = ___ sec/in.

__________________ **2.** Duration of the dark input signal = ___ sec

__________________ **3.** Duration of the light input signal = ___ sec

__________________ **4.** Activating frequency = ___ sec

A photoelectric sensor detects 0.25″ square objects that travel at 150 ft/min and are 0.05″ apart.

_______________ **5.** Speed of the objects = ___ sec/in.

_______________ **6.** Duration of the dark input signal = ___ sec

_______________ **7.** Duration of the light input signal = ___ sec

_______________ **8.** Activating frequency = ___ sec

Vectors and Phase Relationships 13

Name ______________________ Date ____________

True-False

T F **1.** A vector represents the changing value of a cosine wave.

T F **2.** An acute angle is one that is less than 90°.

T F **3.** The vector diagram calculation method connects the heads and tails of the individual vectors.

T F **4.** Vectors can be added only when they are in the same direction.

T F **5.** Mass and length are both examples of scalar quantities.

T F **6.** Mathematical methods often provide answers that are more accurate than graphical methods.

T F **7.** Vectors that are not in phase with each other can still be added directly.

T F **8.** The Greek letter theta (θ) is used to designate an angle.

T F **9.** The parallelogram calculation method is a graphical method.

T F **10.** Vectors cannot be used to represent real-life parameters such as voltage.

Multiple Choice

______________ **1.** A(n) ___ is a vector in which length represents the magnitude of an electrical parameter and direction represents phase angle in electrical degrees.

A. phasor
B. scalar diagram
C. tangent
D. all of the above

______________ **2.** ___ is the branch of mathematics that uses the relationships between the lengths of the sides of a triangle and the angles to perform calculations.
- A. Algebra
- B. Geometry
- C. Trigonometry
- D. Calculus

______________ **3.** The ___ of a right triangle represents the ratio of the lengths of the sides opposite and adjacent to an acute angle.
- A. cosine
- B. sine
- C. cosecant
- D. tangent

______________ **4.** The parallelogram calculation method operates on ___ vectors at a time.
- A. two
- B. three
- C. four
- D. five

______________ **5.** In listing a vector coordinate, the ___-coordinate is listed first.
- A. z
- B. y
- C. x
- D. none of the above

______________ **6.** A(n) ___ is the y-value of a trigonometric function.
- A. abscissa
- B. ordinate
- C. radius vector
- D. angle

Completion

______________ **1.** Vectors represent both direction and ___.

______________ **2.** The ___ is the side of a right triangle that is opposite the right angle.

______________ **3.** Drawing a vector from the tail of the first vector to the head of the last gives a(n) ___ vector.

______________ **4.** A parallelogram is a four-sided plane figure with ___ sides parallel and equal.

______________ **5.** The ___ theorem states that the square of the hypotenuse of a right triangle is equal to the sum of the squares of the other two sides.

_______________ **6.** Coordinates on the x-axis that lie to the ___ of the y-axis are positive.

_______________ **7.** ___ vector solutions are good for providing visual records of calculations.

_______________ **8.** Determining if a trigonometric function is a positive or negative value depends on the ___ location.

Adding Vectors Less Than 90° Apart

Find the resultant vector and angle for adding the following vectors by first calculating the vertical and horizontal components.

Vector A at 6 V and 25°, vector B at 8 V and 35°, and vector C at 5 V and 20°

_______________ **1.** V_A = ___ V

_______________ **2.** V_B = ___ V

_______________ **3.** V_C = ___ V

_______________ **4.** H_A = ___ V

_______________ **5.** H_B = ___ V

_______________ **6.** H_C = ___ V

_______________ **7.** V_{ABC} = ___ V

_______________ **8.** H_{ABC} = ___ V

_______________ **9.** R = ___ V

_______________ **10.** θ = ___°

Vector A at 10 V and 15°, vector B at 12 V and 5°, and vector C at 7 V and 55°

_______________ **11.** V_A = ___ V

_______________ **12.** V_B = ___ V

_______________ **13.** V_C = ___ V

_______________ **14.** H_A = ___ V

_______________ **15.** H_B = ___ V

_______________ **16.** H_C = ___ V

_______________ **17.** V_{ABC} = ___ V

_______________ **18.** H_{ABC} = ___ V

_______________ **19.** R = ___ V

_______________ **20.** θ = ___°

Vector A at 4 V and 65°, vector B at 2 V and 80°, and vector C at 9 V and 10°

________________ **21.** V_A = ___ V

________________ **22.** V_B = ___ V

________________ **23.** V_C = ___ V

________________ **24.** H_A = ___ V

________________ **25.** H_B = ___ V

________________ **26.** H_C = ___ V

________________ **27.** V_{ABC} = ___ V

________________ **28.** H_{ABC} = ___ V

________________ **29.** R = ___ V

________________ **30.** θ = ___°

Subtracting Vectors Less Than 90° Apart

Find the resultant vector angle and magnitude for subtracting the following vectors by first calculating the vertical and horizontal components.

Vector A at 5 V and 50° is subtracted from vector B at 7 V and 35°

________________ **1.** Vector A magnitude = ___ V

________________ **2.** Vector A phase = ___ V

________________ **3.** V_A = ___ V

________________ **4.** H_A = ___ V

________________ **5.** V_B = ___ V

________________ **6.** H_B = ___ V

________________ **7.** V_{B-A} = ___ V

________________ **8.** H_{B-A} = ___ V

________________ **9.** θ_{B-A} = ___°

________________ **10.** Z_{B-A} = ___ V

Vector A at 8 V and 40° is subtracted from vector B at 9 V and 55°

________________ **11.** Vector A magnitude = ___ V

________________ **12.** Vector A phase = ___ V

________________ **13.** V_A = ___ V

________________ **14.** H_A = ___ V

________________ **15.** V_B = ___ V

________________ **16.** H_B = ___ V

________________ **17.** V_{B-A} = ___ V

________________ **18.** H_{B-A} = ___ V

________________ **19.** θ_{B-A} = ___°

________________ **20.** Z_{B-A} = ___ V

Vector A at 6 V and 25° is subtracted from vector B at 3 V and 75°

________________ **21.** Vector A magnitude = ___ V

________________ **22.** Vector A phase = ___ V

________________ **23.** V_A = ___ V

________________ **24.** H_A = ___ V

________________ **25.** V_B = ___ V

________________ **26.** H_B = ___ V

________________ **27.** V_{B-A} = ___ V

________________ **28.** H_{B-A} = ___ V

________________ **29.** θ_{B-A} = ___°

________________ **30.** Z_{B-A} = ___ V

Phase Relationships

Consider an electrical motor with a series capacitive AC circuit. In the circuit, the source voltage lags behind the line current by 90°. This can be shown in phasor and vector graph form.

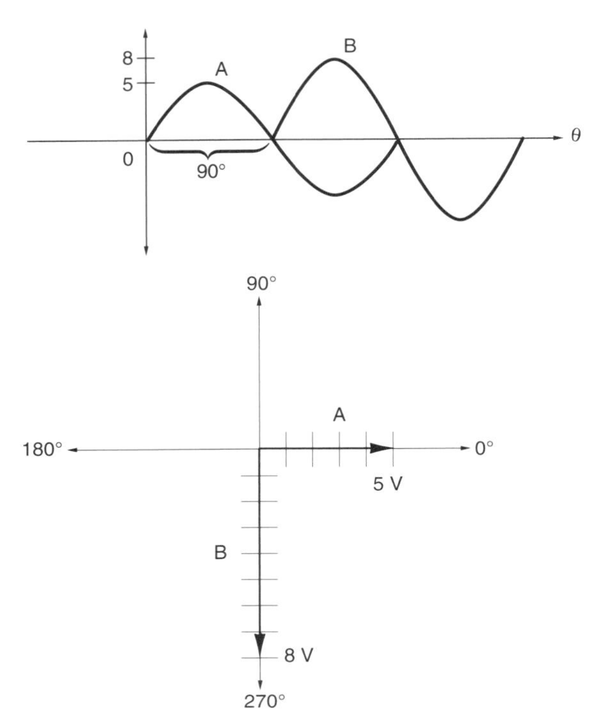

Solve for the resultant vector and angle theta by using the vector diagram calculation method.

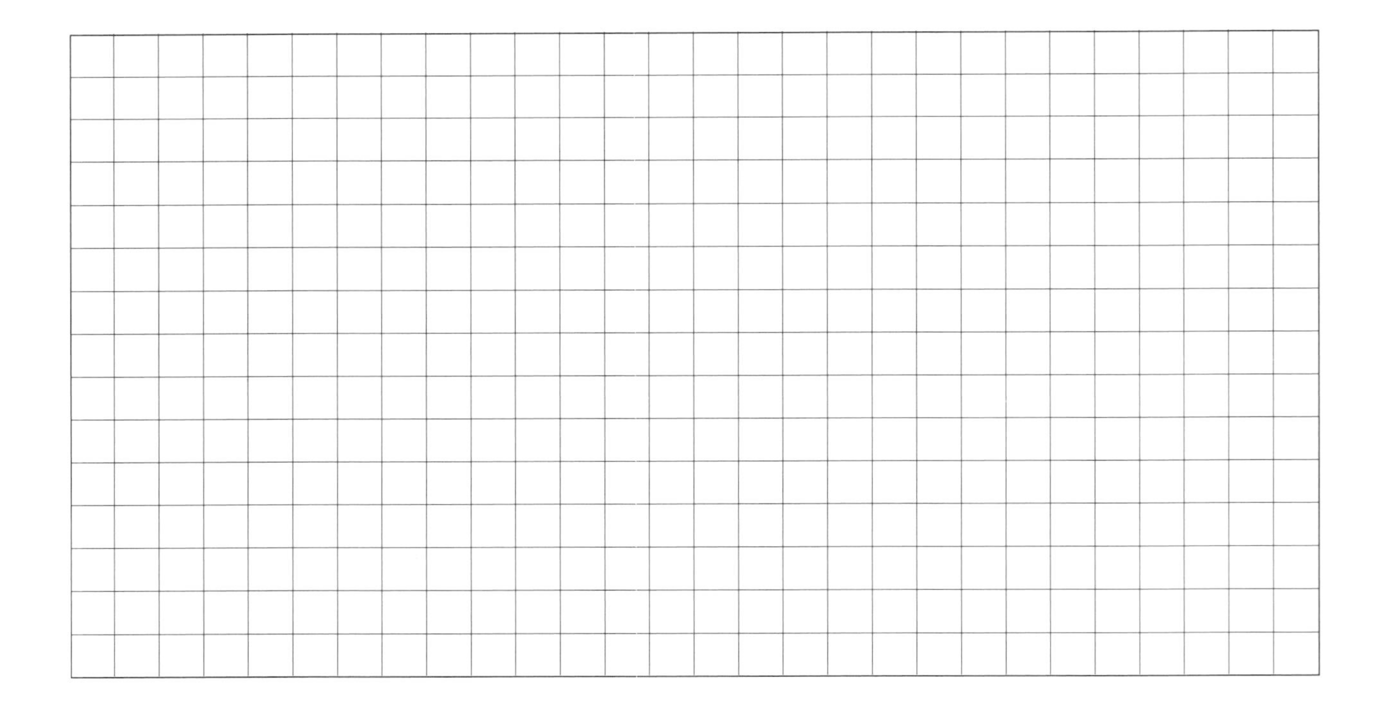

_______________ **1.** V_R = ___ V

_______________ **2.** θ = ___°

Now solve for the resultant vector and angle theta by using the Pythagorean theorem.

_______________ **3.** V_R = ___ V

_______________ **4.** θ = ___°

Resistive AC Circuits 14

Name ________________________________ Date ____________

True-False

T F **1.** The common value of voltage used in homes in the United States is 240 VAC.

T F **2.** Power dissipated in a circuit is always negative.

T F **3.** An average DC current causes greater heat than an average AC current.

T F **4.** Bleeder resistors reduce changes in the ripple voltage when current is delivered to the load.

T F **5.** When an AC value is given without it being specified, it is assumed to be a peak value.

T F **6.** Peak power is equal to effective voltage multiplied by effective current.

T F **7.** A peak-to-peak value measures from the maximum positive to the maximum negative alternation of a sine wave.

T F **8.** An average value refers to a whole cycle of a sine wave.

T F **9.** Total current is equal to current flow through any component in a series resistive circuit.

T F **10.** Resistances connected in series are subtracted from each other in an AC circuit.

Multiple Choice

______________ **1.** The ___ value of AC voltage is the value of a sine wave that produces the same amount of heat as DC of the same value.

A. peak
B. effective
C. peak-to-peak
D. average

______________ **2.** A ___ is a network of resistances, capacitances, and inductances that provides little opposition to certain frequencies while blocking others.
A. filter
B. ripple
C. bleeder
D. none of the above

______________ **3.** A ___ curve is a graph that shows the changes in circuit power as AC parameters change over time.
A. sine
B. wave
C. voltage
D. power

______________ **4.** In a conductor rotating through 360°, maximum positive voltage is inducted at ___° rotation.
A. 0
B. 90
C. 180
D. 270

______________ **5.** The average value of an AC voltage or current is the ___ of all instantaneous values in a sine wave.
A. total
B. addition
C. mean
D. subtraction

Completion

______________ **1.** The ___ value of an AC voltage or current refers to the maximum instantaneous value of the sine wave.

______________ **2.** Ripple voltage is output voltage of a power source harmonically related in ___ to the power source.

______________ **3.** Bleeder resistors are used to "bleed" residual charges off of a(n) ___ when the power source is turned OFF.

______________ **4.** In a pure resistive ___ circuit, the current sine wave and voltage sine wave are in phase with each other.

______________ **5.** Power is positive and reaches its ___ value at 90° and 270°.

______________ **6.** ___ capacitors are used to control harmonics problems in the power source.

Calculating Power in an AC Resistive Circuit

Calculate the peak voltage, current, and power in the following circuits:

A circuit where effective voltage is 130 V and effective current is 4 A

_______________ **1.** V_P = ___ V

_______________ **2.** I_P = ___ A

_______________ **3.** P_P = ___ W

A circuit where effective voltage is 120 V and effective current is 8 A

_______________ **4.** V_P = ___ V

_______________ **5.** I_P = ___ A

_______________ **6.** P_P = ___ W

A circuit where effective voltage is 115 V and effective current is 6 A

_______________ **7.** V_P = ___ V

_______________ **8.** I_P = ___ A

_______________ **9.** P_P = ___ W

Calculate the average power in the following circuits:

A circuit with a peak voltage of 180 V and peak current of 9 A

_______________ **10.** P_{AVG} = ___ W

A circuit with a peak voltage of 240 V and peak current of 6 A

_______________ **11.** P_{AVG} = ___ W

A circuit with a peak voltage of 230 V and peak current of 7 A

_______________ **12.** P_{AVG} = ___ W

Calculating Power Dissipated in a Combination Circuit

Consider a 130 VAC series/parallel circuit that has a series resistance of 5 Ω and parallel resistances of 12 Ω and 20 Ω.

Calculate total series resistance, total parallel resistance, and total circuit resistance.

_______________ **1.** R_{ST} = ___ Ω

_______________ **2.** R_{PT} = ___ Ω

_______________ **3.** R_{T} = ___ Ω

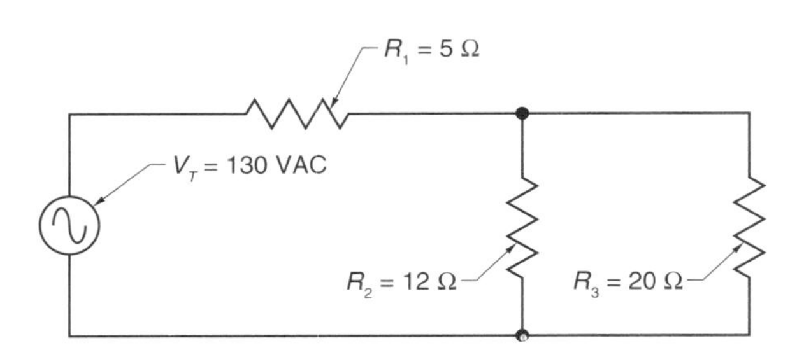

Calculate total current, voltage drop across the circuit, current across the loads in parallel, and total current.

_______________ **4.** I_{T} = ___ A

_______________ **5.** V_{D1} = ___ V

_______________ **6.** $V_{D2,3}$ = ___ V

_______________ **7.** I_{2} = ___ A

_______________ **8.** I_{3} = ___ A

_______________ **9.** I_{T} = ___ A

Verify calculations using Kirchhoff's voltage and current laws.

_______________ **10.** V_{N} = ___ V

_______________ **11.** I_{N} = ___ A

Calculate total power consumed and power dissipated in each load.

_______________ **12.** P_{T} = ___ W

_______________ **13.** P_{D1} = ___ W

_______________ **14.** P_{D2} = ___ W

_______________ **15.** P_{D3} = ___ W

Series Resistive AC Circuits

Calculate the total resistance, total current, and voltage drops across each resistor.

_______________ **1.** R_T = ___ Ω

_______________ **2.** I_T = ___ A

_______________ **3.** V_{D1} = ___ V

_______________ **4.** V_{D2} = ___ V

_______________ **5.** V_{D3} = ___ V

_______________ **6.** V_{D4} = ___ V

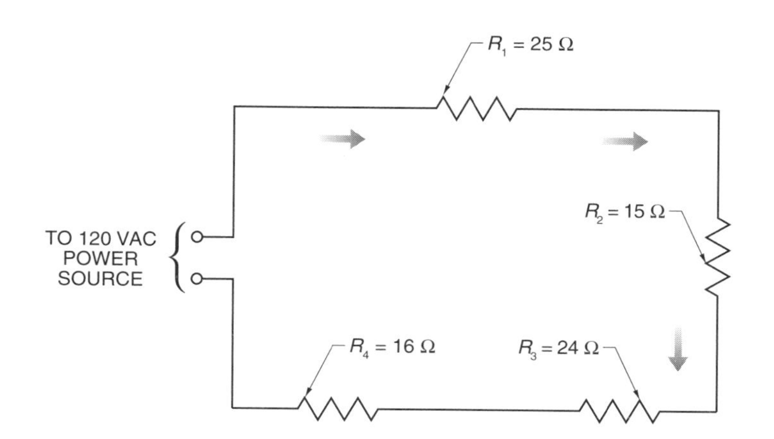

Calculate the total power dissipated in the circuit.

_______________ **7.** P_D = ___ W

Calculate the total resistance, total current, and voltage drops across each resistor.

________________ **8.** R_T = ___ Ω

________________ **9.** I_T = ___ A

________________ **10.** V_{D1} = ___ V

________________ **11.** V_{D2} = ___ V

________________ **12.** V_{D3} = ___ V

________________ **13.** V_{D4} = ___ V

________________ **14.** V_{D5} = ___ V

________________ **15.** V_{D6} = ___ V

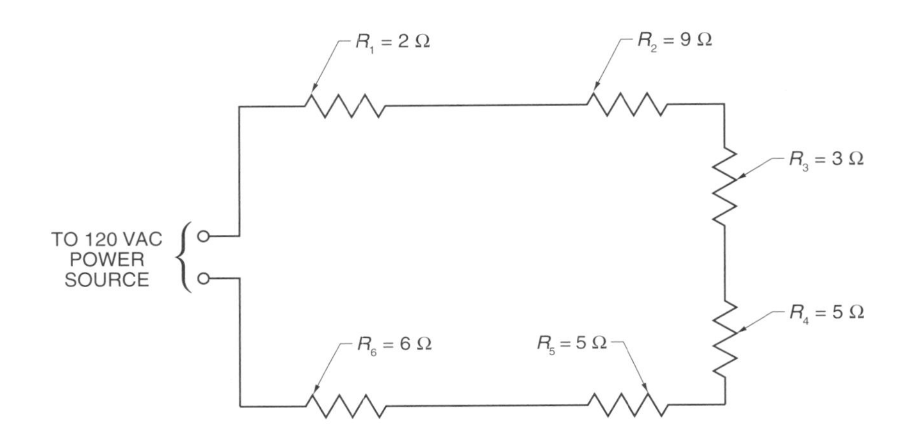

Calculate the total power dissipated in the circuit.

________________ **16.** P_D = ___ W

Parallel Resistive AC Circuits

Calculate each value of resistance and the total circuit resistance.

__________________ **1.** R_1 = ___ Ω

__________________ **2.** R_2 = ___ Ω

__________________ **3.** R_3 = ___ Ω

__________________ **4.** R_T = ___ Ω

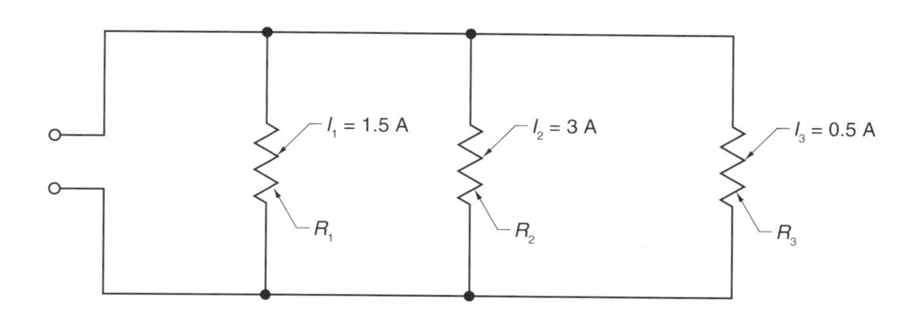

Calculate the total power dissipated in the circuit.

__________________ **5.** P_D = ___ W

Calculate each value of resistance and the total circuit resistance.

__________________ **6.** R_1 = ___ Ω

__________________ **7.** R_2 = ___ Ω

__________________ **8.** R_3 = ___ Ω

__________________ **9.** R_4 = ___ Ω

__________________ **10.** R_T = ___ Ω

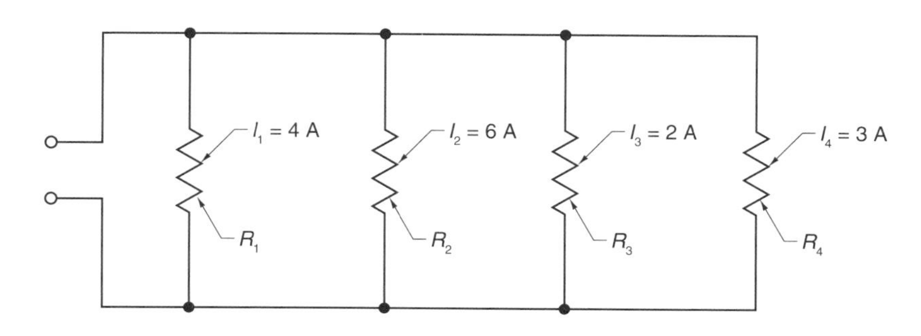

Calculate the total power dissipated in the circuit.

__________________ **11.** P_D = ___ W

Inductive AC Circuits 15

Name ______________________________ Date ______________

True-False

T F **1.** A pure inductive circuit is an AC circuit with no resistance.

T F **2.** Kirchhoff's voltage law states that the algebraic sum of the voltage in a closed-loop circuit is equal to 0.

T F **3.** Current in an inductive AC circuit can be calculated without knowing source voltage.

T F **4.** Apparent power represents a load that has both true power and reactive power.

T F **5.** Inductors have no resistance.

T F **6.** Impedance is measured in joules.

T F **7.** Poor power factor exists when the power factor in a circuit is less than 90%.

T F **8.** As frequency increases in a series inductive-resistive AC circuit, both impedance and current increase as well.

T F **9.** In inductive-resistive AC circuits, voltage is the same across all components connected in parallel.

T F **10.** Frequency does not affect resistance in a parallel inductive-resistive AC circuit.

T F **11.** In an inductive AC circuit, counter-electromotive force is 180° out of phase with the source voltage.

T F **12.** Impedance aids the flow of alternating current.

T F **13.** True power is measured in volt-amperes and apparent power is measured in watts.

T F **14.** Only resistance can dissipate power in an electrical circuit.

T F **15.** The standard power source line frequency in Canada is 120 Hz.

Multiple Choice

_______________ **1.** ___ is the ability of a reactive circuit to conduct current and is measured in siemens.
A. Induction
B. Reactance
C. Susceptance
D. Reactive power

_______________ **2.** Inductive reactance is affected by the size of the inductor and ___.
A. AC frequency
B. DC frequency
C. power
D. voltage

_______________ **3.** The ability of a circuit containing both resistance and reactance to conduct current is known as ___.
A. impedance
B. admittance
C. reactance
D. none of the above

_______________ **4.** In an inductive-resistive AC circuit, almost all of the power is ___ power.
A. nonexistent
B. apparent
C. true
D. reactive

_______________ **5.** Under normal conditions in power circuits, ___ power factor is the most desirable condition.
A. zero
B. unity
C. poor
D. all of the above

Completion

_______________ **1.** Inductive ___ is the opposition of an inductor to alternating current.

_______________ **2.** ___ factor is the ratio of the inductive reactance to the resistance of an inductor.

_______________ **3.** A voltage ___ is a vector diagram used to show the magnitude and phase of voltage in a series inductive-resistive AC circuit.

__________________ **4.** Impedance in an AC circuit can be found in the form of capacitive reactance or ___ reactance.

__________________ **5.** Power ___ is the ratio of true power to apparent power.

__________________ **6.** ___ is used to calculate most parameters in a parallel inductive-resistive circuit.

Calculating Inductive Reactance

Calculate the inductive reactance in the following circuits:

An inductive AC circuit with a 60 Hz lamp, an inductance value of 0.5 H, and a voltage of 100 V

__________________ **1.** X_L = ___ Ω

__________________ **2.** I_L = ___ A

An inductive AC circuit with a 50 Hz lamp, an inductance value of 0.8 H, and a voltage of 120 V

__________________ **3.** X_L = ___ Ω

__________________ **4.** I_L = ___ A

An inductive AC circuit with a 45 Hz lamp, an inductance value of 0.2 H, and a voltage of 130 V

__________________ **5.** X_L = ___ Ω

__________________ **6.** I_L = ___ A

Calculating Resistance and Impedance

Calculate the total resistance and impedance in the following circuits:

A 130 VAC inductive circuit with an 80 W lamp and an inductive reactance of 160 Ω

__________________ **1.** R_T = ___ Ω

__________________ **2.** Z_T = ___ Ω

A 140 VAC inductive circuit with a 100 W lamp and an inductive reactance of 170 Ω

__________________ **3.** R_T = ___ Ω

__________________ **4.** Z_T = ___ Ω

A 120 VAC inductive circuit with a 75 W lamp and an inductive reactance of 150 Ω

__________________ **5.** R_T = ___ Ω

__________________ **6.** Z_T = ___ Ω

Calculating Source Voltage

Calculate the source voltage for the following series inductive-resistive AC circuits:

A 130 VAC lamp and an inductor with a voltage drop of 210 V

________________ **1.** V_A = ___ V

A 120 VAC lamp and an inductor with a voltage drop of 240 V

________________ **2.** V_A = ___ V

A 150 VAC lamp and an inductor with a voltage drop of 140 V

________________ **3.** V_A = ___ V

Impedance in Series Inductive-Resistive Circuits

Calculate the component resistance, circuit inductive reactance, and circuit impedance for the following series inductive-resistive AC circuits:

A circuit with a resistive current of 0.85 A, 130 V across a lamp, and 208 V across an inductor

________________ **1.** R_N = ___ Ω

________________ **2.** X_L = ___ Ω

________________ **3.** Z = ___ Ω

A circuit with a resistive current of 0.93 A, 120 V across a lamp, and 216 V across an inductor

________________ **4.** R_N = ___ Ω

________________ **5.** X_L = ___ Ω

________________ **6.** Z = ___ Ω

A circuit with a resistive current of 0.83 A, 140 V across a lamp, and 208 V across an inductor

________________ **7.** R_N = ___ Ω

________________ **8.** X_L = ___ Ω

________________ **9.** Z = ___ Ω

Inductive AC Circuits

Consider a 120 VAC circuit with a frequency of 60 Hz, an inductance of 0.8 H, and a 75 W lamp.

Draw the circuit in the space provided below using the correct symbols. Label each component.

Using the circuit information, solve for inductive reactance, component resistance, and impedance.

__________________ **1.** X_L = ___ Ω

__________________ **2.** R_T = ___ Ω

__________________ **3.** Z_T = ___ Ω

Calculate total circuit current, apparent power, and power dissipated.

__________________ **4.** I_T = ___ A

__________________ **5.** P_A = ___ W

__________________ **6.** P_D = ___ W

Inductive Reactance

Calculate the following:

____________________ **1.** X_L = ___ Ω

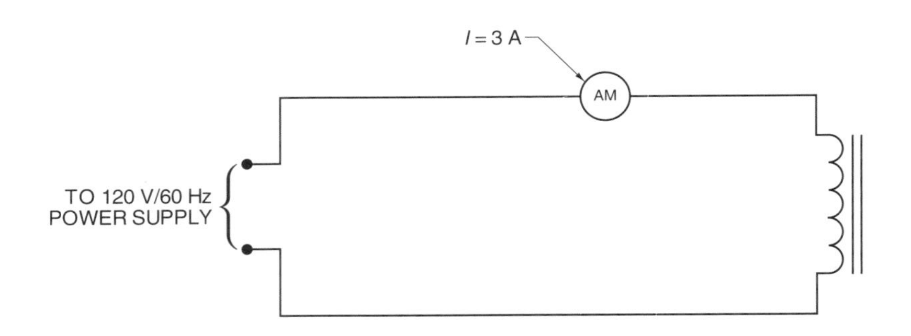

____________________ **2.** X_L = ___ Ω

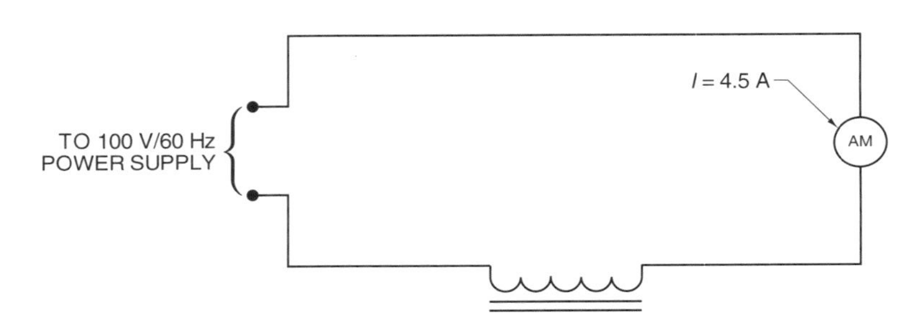

____________________ **3.** X_L = ___ Ω

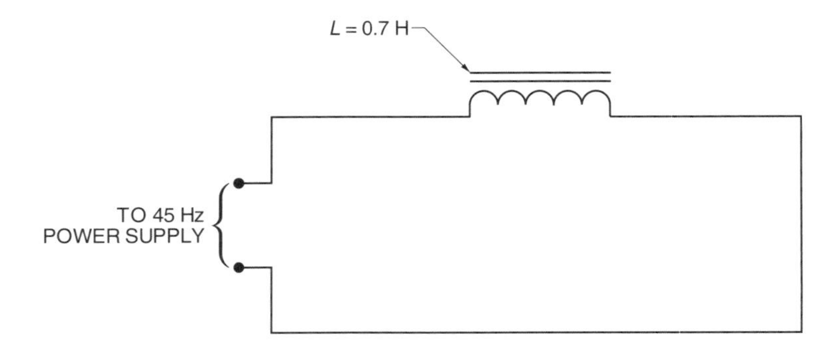

4. X_L = ___ Ω

______________________ **5.** L_T = ___ H

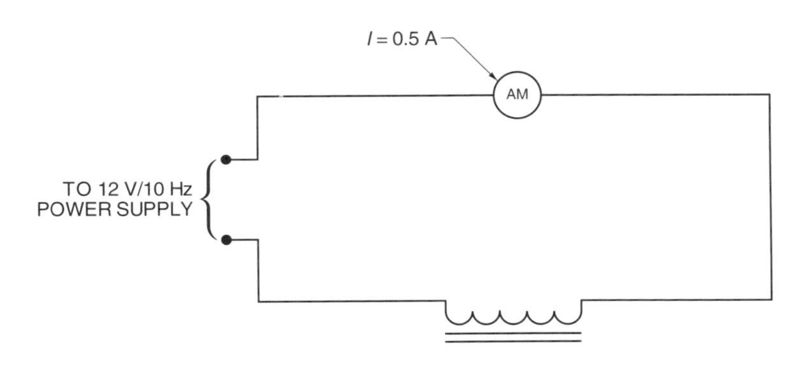

______________________ **6.** L_T = ___ H

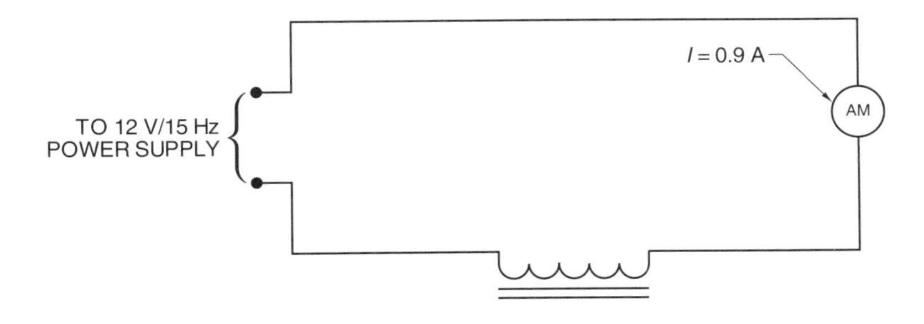

______________ **7.** f = ___ Hz

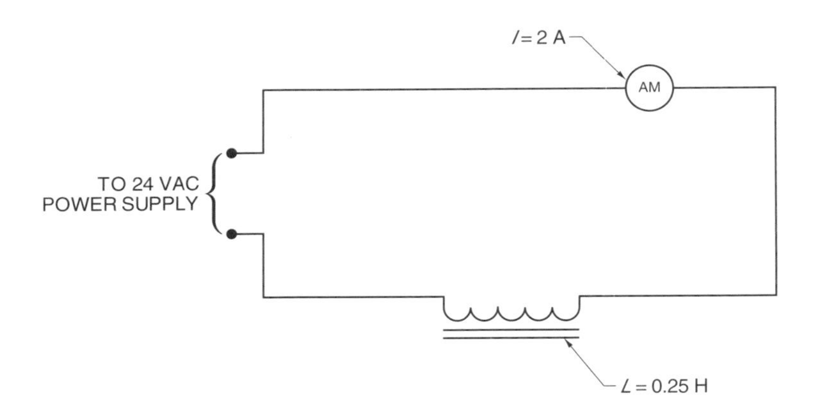

I = 2 A
AM
TO 24 VAC
POWER SUPPLY
L = 0.25 H

______________ **8.** f = ___ Hz

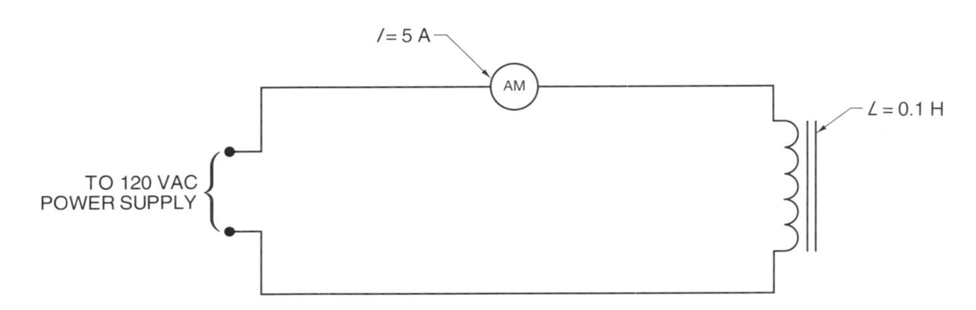

I = 5 A
AM
L = 0.1 H
TO 120 VAC
POWER SUPPLY

Capacitive AC Circuits 16

Name ______________________________ Date ______________

True-False

T F **1.** A pure capacitive circuit is theoretical because it involves no resistance.

T F **2.** Adding capacitance to an AC series circuit increases the circuit current.

T F **3.** In a simple circuit, current is the reference point.

T F **4.** In a capacitor, consumed power has a positive value, while power returned to a circuit is negative.

T F **5.** Frequency does not affect capacitive reactance.

T F **6.** Angle theta is the angle by which current leads the source voltage.

T F **7.** As capacitive reactance increases, circuit impedance increases.

T F **8.** Resistive current is in phase with source voltage in a parallel resistive-capacitive circuit.

T F **9.** The Pythagorean theorem can be used to calculate the source current in a parallel resistive-capacitive circuit.

T F **10.** Series/parallel resistive-capacitive circuits should not be used for electronics applications.

T F **11.** Power in a pure capacitive circuit is the equivalent of reactive power in a capacitive circuit.

T F **12.** In a pure capacitive circuit, the capacitor uses energy to charge, and half of this energy is returned to the circuit when the capacitor discharges.

T F **13.** In a series resistive-capacitive circuit, impedance is the combination of inductance and capacitive reactance to oppose the flow of voltage.

T F **14.** In an inductive-resistive circuit, apparent power is lagging, indicating inductance.

T F **15.** A change in impedance also causes a change in capacitive reactance.

Multiple Choice

______ **1.** ___ reactance is the opposition to current flow by a capacitor.
A. Inductive
B. Impedance
C. Capacitive
D. none of the above

______ **2.** Capacitive reactance is measured in ___.
A. ohms
B. volts
C. watts
D. volt-amperes

______ **3.** In a pure capacitive AC circuit, which of the following values is needed in order to calculate other parameters?
A. frequency
B. capacitance
C. voltage
D. all of the above

______ **4.** In a parallel resistive-capacitive circuit, frequency has no effect on source voltage or ___.
A. capacitive reactance
B. reactive power
C. resistance
D. all of the above

______ **5.** In a series/parallel resistive-capacitive circuit, ___ is used as the reference point for the series components.
A. current
B. voltage
C. resistance
D. capacitance

______ **6.** In a series/parallel resistive-capacitive circuit, ___ is used as the reference point for the parallel branches.
A. current
B. voltage
C. resistance
D. capacitance

Completion

______________ **1.** Capacitive voltage is equal to the ___ voltage at all times in a closed-loop circuit.

______________ **2.** If frequency is changed in a series resistive-capacitive circuit, the only parameter directly affected is ___.

______________ **3.** Dividing source voltage by the total circuit current yields ___.

______________ **4.** A change in capacitive reactance results in different capacitive ___ in each branch circuit of a parallel resistive-capacitive circuit.

______________ **5.** Capacitive reactance is ___ proportional to frequency.

______________ **6.** In any series resistive-capacitive circuit, the resistive current and the ___ across the resistor are in phase.

______________ **7.** In a series resistive-capacitive circuit, only ___ can dissipate true power.

______________ **8.** In a series resistive-capacitive circuit, ___ leads current and source voltage, indicating capacitance.

______________ **9.** Increasing the amount of ___ in series resistive-capacitive circuits reduces the amount of capacitive reactance.

______________ **10.** ___ is the same across all components connected in parallel in resistive-capacitive circuits.

Calculating Current and Capacitive Reactance

Calculate the capacitive reactance and total current for the following circuits:

A pure capacitive circuit that has a 10 μF capacitance connected to a 240 VAC/60 Hz power source

______________ **1.** X_C = ___ Ω

______________ **2.** I_T = ___ A

A pure capacitive circuit that has a 12 μF capacitance connected to a 220 VAC/60 Hz power source

______________ **3.** X_C = ___ Ω

______________ **4.** I_T = ___ A

A pure capacitive circuit that has a 15 μF capacitance connected to a 230 VAC/50 Hz power source

______________ **5.** X_C = ___ Ω

______________ **6.** I_T = ___ A

Series Resistive-Capacitive Circuit Calculations

Calculate the angle theta and source voltage for the following circuits:

A series resistive-capacitive circuit with resistive voltage of 208 V and capacitive voltage of 130 V

________________ **1.** θ = ___°

________________ **2.** V_A = ___ V

A series resistive-capacitive circuit with resistive voltage of 210 V and capacitive voltage of 120 V

________________ **3.** θ = ___°

________________ **4.** V_A = ___ V

A series resistive-capacitive circuit with resistive voltage of 220 V and capacitive voltage of 150 V

________________ **5.** θ = ___°

________________ **6.** V_A = ___ V

Parallel Resistive-Capacitive Circuit Calculations

Calculate the angle theta and source current for the following circuits:

A parallel resistive-capacitive circuit with resistive current of 8 A and capacitive current of 10 A

________________ **1.** θ = ___°

________________ **2.** I_T = ___ V

A parallel resistive-capacitive circuit with resistive current of 6 A and capacitive current of 10 A

________________ **3.** θ = ___°

________________ **4.** I_T = ___ V

A parallel resistive-capacitive circuit with resistive current of 4 A and capacitive current of 8 A

________________ **5.** θ = ___°

________________ **6.** I_T = ___ V

Capacitive Reactance—Series Circuits

Calculate the total capacitance and capacitive reactance for each of the following circuits:

__________________ **1.** C_T = ___ μF

__________________ **2.** X_C = ___ Ω

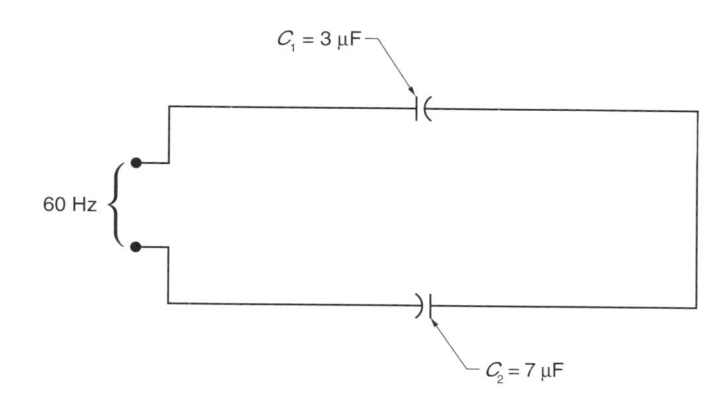

__________________ **3.** C_T = ___ μF

__________________ **4.** X_C = ___ Ω

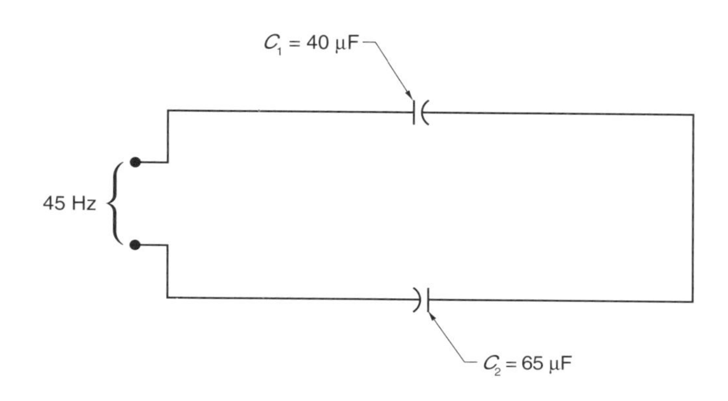

__________________ **5.** C_T = ___ μF

__________________ **6.** X_C = ___ Ω

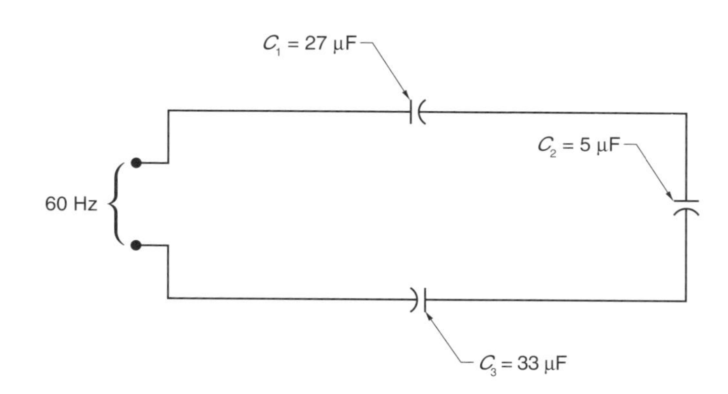

_______________ **7.** C_T = ___ μF

_______________ **8.** X_C = ___ Ω

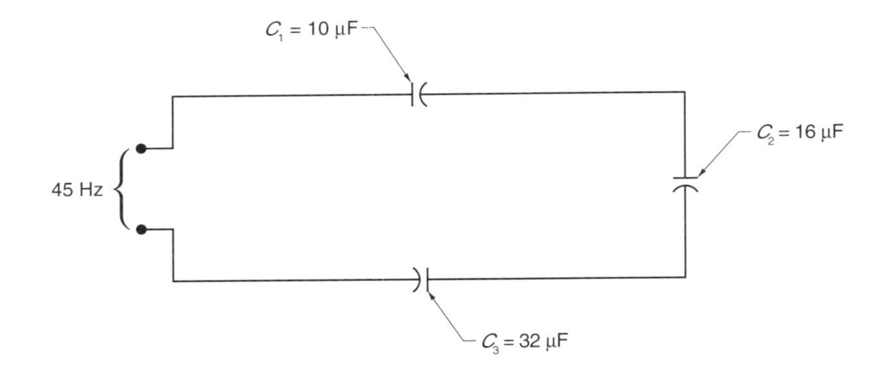

_______________ **9.** C_T = ___ μF

_______________ **10.** X_C = ___ Ω

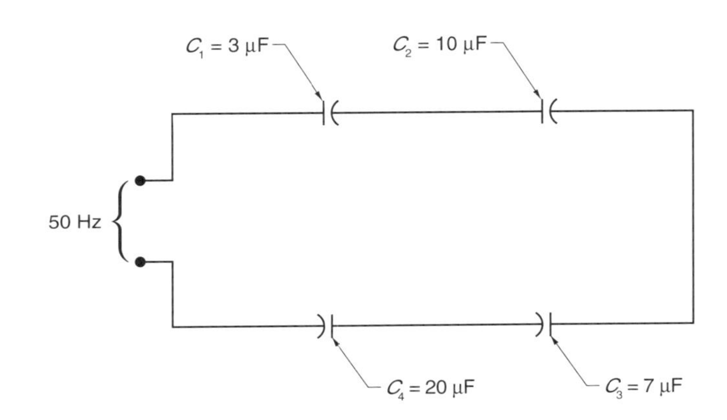

_______________ **11.** C_T = ___ μF

_______________ **12.** X_C = ___ Ω

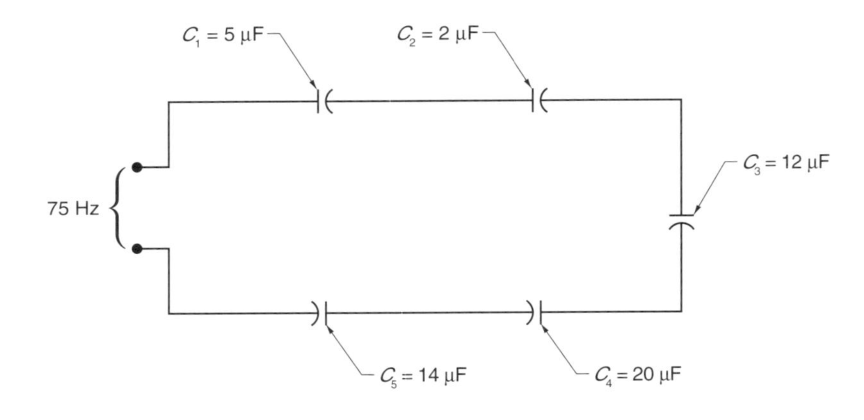

Capacitive Reactance—Parallel Circuits

Calculate the total capacitance and capacitive reactance for each of the following circuits:

____________________ **1.** C_T = ___ μF

____________________ **2.** X_C = ___ Ω

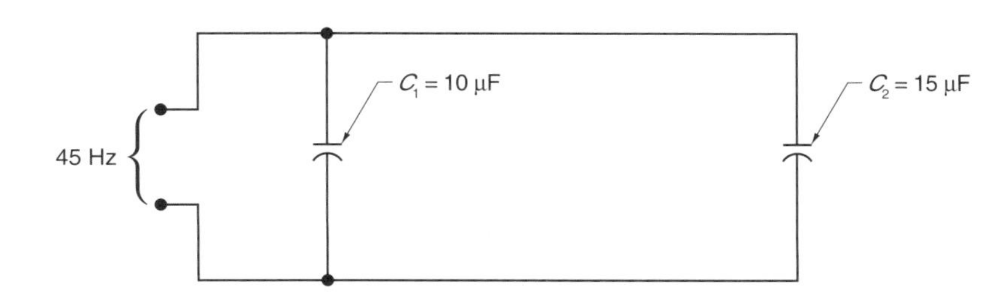

____________________ **3.** C_T = ___ μF

____________________ **4.** X_C = ___ Ω

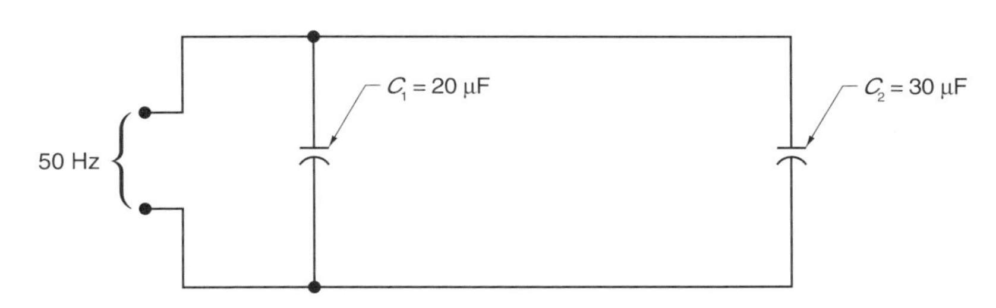

____________________ **5.** C_T = ___ μF

____________________ **6.** X_C = ___ Ω

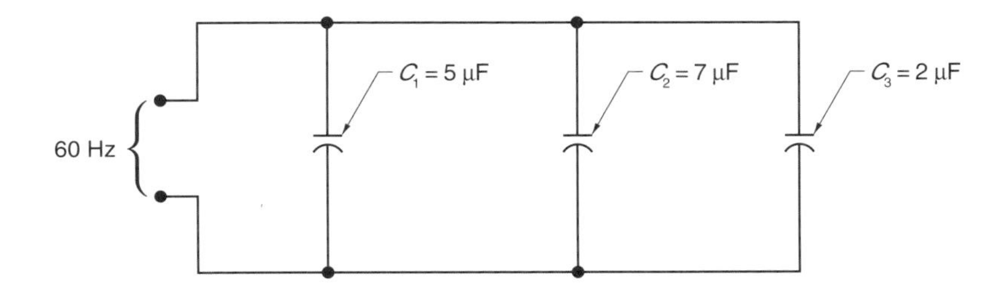

________________ **7.** C_T = ___ μF

________________ **8.** X_C = ___ Ω

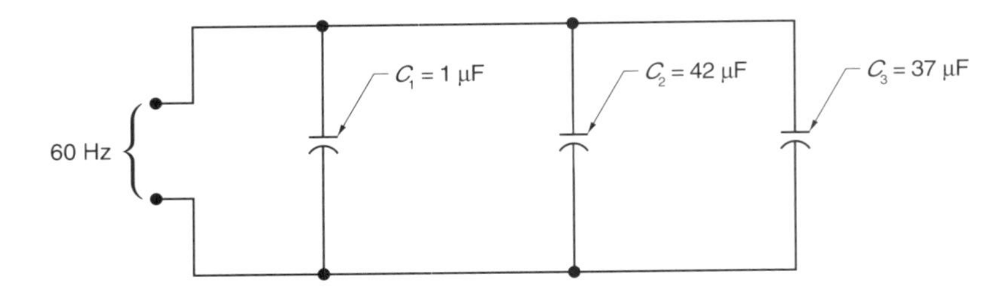

________________ **9.** C_T = ___ μF

________________ **10.** X_C = ___ Ω

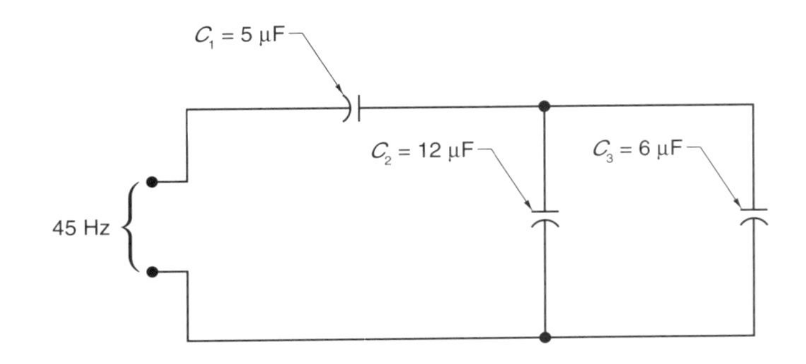

________________ **11.** C_T = ___ μF

________________ **12.** X_C = ___ Ω

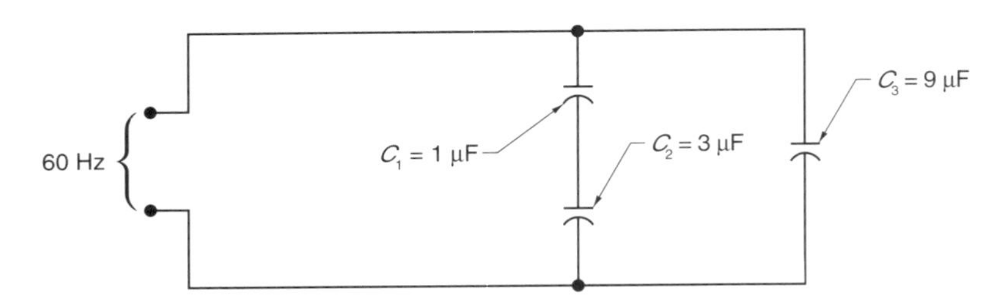

Inductive-Resistive-Capacitive Circuits 17

Name ______________________ Date ____________

True-False

T F **1.** When a series circuit is inductive, the current lags the source voltage.

T F **2.** Increasing the frequency in a parallel inductive-resistive-capacitive circuit causes current flow to increase in the inductive branch.

T F **3.** When a series circuit is capacitive, line current leads the source voltage.

T F **4.** Total reactive power can be calculated by subtracting the inductive power vector from the capacitive power vector.

T F **5.** Inductive and capacitive branches of a circuit have no resistance.

T F **6.** The resistive branch circuit only opposes current flow.

T F **7.** Current is often reduced to its vertical and horizontal components in order to calculate total line current.

T F **8.** When a load is a series resistive-capacitive combination, current flow is initially at its minimum value.

T F **9.** In a series resistive AC circuit, source voltage and line current are in phase with each other.

T F **10.** In a series inductive-resistive-capacitive circuit, inductive and capacitive voltages are in phase with each other.

T F **11.** Current is used as the reference in a parallel inductive-resistive-capacitive circuit.

T F **12.** In a parallel inductive-resistive-capacitive circuit, the power factor is equal to the true power divided by the apparent power.

T F **13.** Frequency must be known in order to calculate inductive reactance.

T F **14.** In a parallel inductive-resistive-capacitive circuit, a leading current indicates an inductive circuit.

Multiple Choice

_______________ **1.** A power ___ diagram is used to calculate apparent power in a parallel inductive-resistive-capacitive circuit.
- A. vector
- B. angle
- C. phase
- D. degree

_______________ **2.** Series/parallel inductive-resistive-capacitive circuits are usually analyzed to calculate all of the following except ___.
- A. power factor
- B. circuit power
- C. frequency
- D. total circuit current

_______________ **3.** ___ is the same across all branch circuits of a parallel inductive-resistive-capacitive circuit.
- A. Current
- B. Voltage
- C. Resistance
- D. Capacitance

_______________ **4.** ___ is the same throughout the circuit of a series inductive-resistive-capacitive circuit.
- A. Voltage
- B. Capacitance
- C. Inductance
- D. Current

_______________ **5.** In a series inductive-resistive-capacitive circuit, circuit current is equal to source voltage divided by circuit ___.
- A. voltage
- B. current
- C. resistance
- D. impedance

_______________ **6.** Without taking any measurements, all of the following values are typically known in a series inductive-resistive-capacitive circuit except ___.
- A. frequency
- B. source voltage
- C. capacitance
- D. none of the above

Completion

________________ **1.** A circuit where all inductive, resistive, and capacitive circuit elements are connected in one current path is known as a series ___ circuit.

________________ **2.** In a series circuit, if inductive reactance is larger than capacitive reactance, then the circuit is ___.

________________ **3.** Resistive branch circuit current is in phase with resistive branch circuit ___.

________________ **4.** The vector diagram calculation method or the ___ can be used to calculate total current in a series/parallel inductive-resistive-capacitive circuit.

________________ **5.** With an inductive or capacitive circuit, inductance and capacitance present an opposition to current flow known as ___.

Series Inductive-Resistive-Capacitive Circuit Calculations

Analyze the circuit in the following steps.

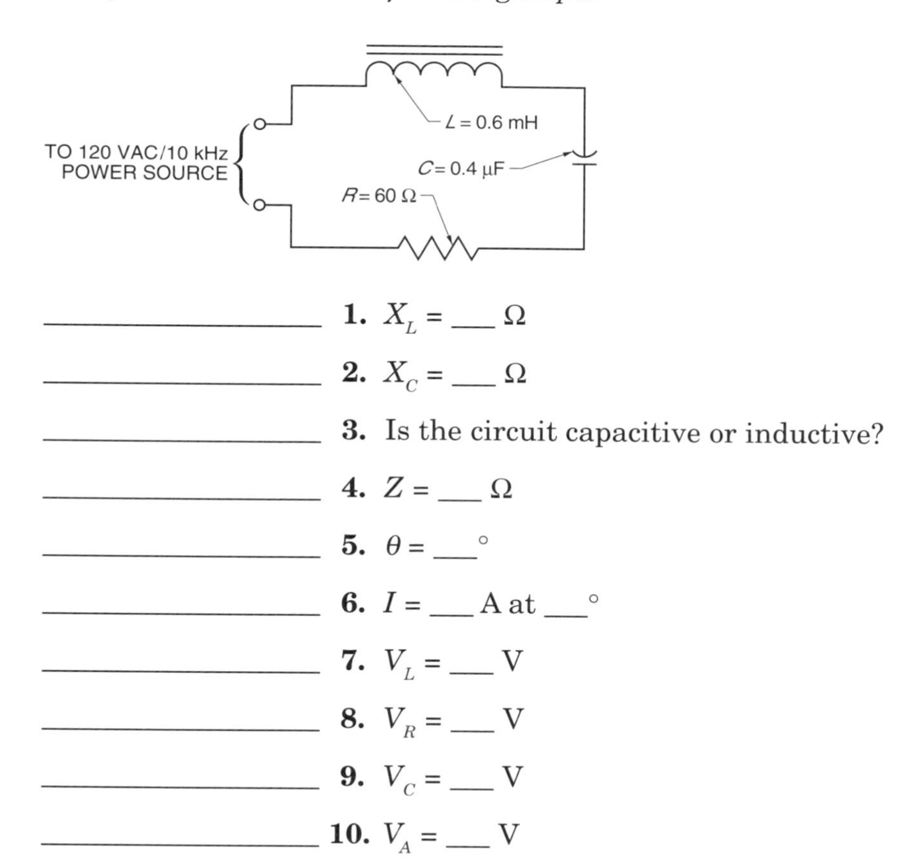

________________ **1.** X_L = ___ Ω

________________ **2.** X_C = ___ Ω

________________ **3.** Is the circuit capacitive or inductive?

________________ **4.** Z = ___ Ω

________________ **5.** θ = ___°

________________ **6.** I = ___ A at ___°

________________ **7.** V_L = ___ V

________________ **8.** V_R = ___ V

________________ **9.** V_C = ___ V

________________ **10.** V_A = ___ V

________________ **11.** P_{APP} = ___ VA

________________ **12.** $P_{VAR\text{-}L}$ = ___ VAR

________________ **13.** $P_{VAR\text{-}C}$ = ___ VAR

________________ **14.** P_{TRUE} = ___ W

________________ **15.** PF = ___%

Parallel Inductive-Resistive-Capacitive Circuit Calculations

Analyze the circuit in the following steps.

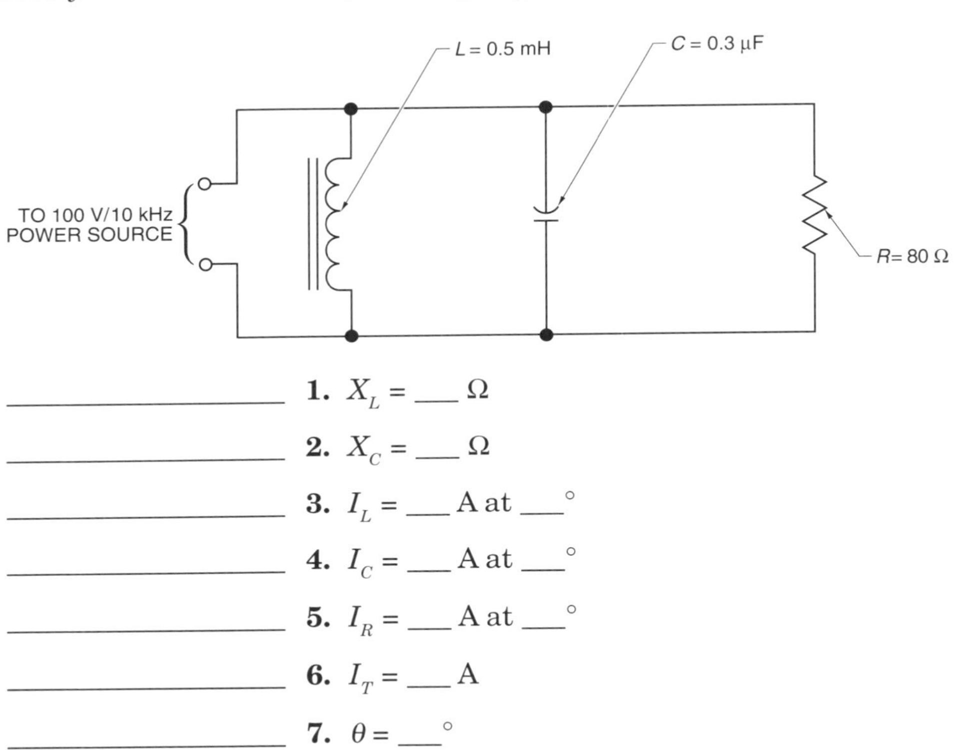

________________ **1.** X_L = ___ Ω

________________ **2.** X_C = ___ Ω

________________ **3.** I_L = ___ A at ___°

________________ **4.** I_C = ___ A at ___°

________________ **5.** I_R = ___ A at ___°

________________ **6.** I_T = ___ A

________________ **7.** θ = ___°

________________ **8.** Z = ___ Ω at ___°

________________ **9.** P_{APP} = ___ VA

________________ **10.** P_{TRUE} = ___ W

________________ **11.** P_{VAR} = ___ VAR

________________ **12.** PF = ___%

Series/Parallel Inductive-Resistive-Capacitive Circuit Calculations

Analyze the circuit in the following steps.

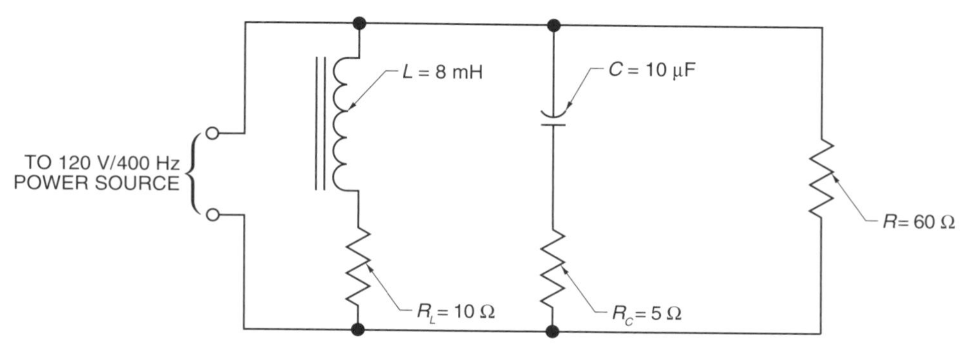

________________ **1.** X_L = ___ Ω

________________ **2.** X_C = ___ Ω

________________ **3.** θ_L = ___°

________________ **4.** θ_C = ___°

________________ **5.** Z_L = ___ Ω at ___°

________________ **6.** Z_C = ___ Ω at ___°

________________ **7.** I_L = ___ A at ___°

________________ **8.** I_C = ___ A at ___°

________________ **9.** I_R = ___ A at ___°

________________ **10.** V_L = ___ A

________________ **11.** H_L = ___ A

________________ **12.** V_C = ___ A

________________ **13.** H_C = ___ A

________________ **14.** V_T = ___ A

________________ **15.** H_T = ___ A

________________ **16.** θ = ___°

________________ **17.** I_T = ___ A at ___°

________________ **18.** Z_T = ___ Ω at ___°

________________ **19.** P_{APP} = ___ VA

________________ **20.** P_{TRUE} = ___ W

________________ **21.** P_X = ___ VAR

Constructing a Series Inductive-Resistive-Capacitive Circuit

Consider a series inductive-resistive-capacitive circuit with a frequency of 12 kHz and a source voltage of 140 VAC. The inductive reactance is 15 Ω, the capacitive reactance is 19 Ω, and the impedance is 30 Ω.

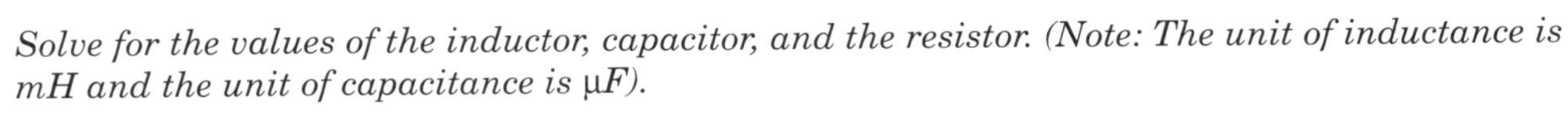

Solve for the values of the inductor, capacitor, and the resistor. (Note: The unit of inductance is mH and the unit of capacitance is μF).

____________________ **1.** L = ___ mH

____________________ **2.** C = ___ μF

____________________ **3.** R = ___ Ω

Draw the circuit using the correct symbols and include all values on the drawing.

Constructing a Parallel Inductive-Resistive-Capacitive Circuit

Consider a parallel inductive-resistive-capacitive circuit with a frequency of 10 kHz and a source voltage of 115 VAC. The inductive reactance is 25 Ω, the capacitive reactance is 32 Ω, and the resistive branch circuit current is 2.5 A.

Solve for the values of the inductor, capacitor, and the resistor. (Note: The unit of inductance is mH and the unit of capacitance is μF.)

____________________ **1.** L = ___ mH

____________________ **2.** C = ___ μF

____________________ **3.** R = ___ Ω

Draw the circuit using the correct symbols and include all values on the drawing.

Resonance 18

Name ______________________________ Date ______________

True-False

T F **1.** The angle theta of a series resonant circuit is equal to 0°.

T F **2.** Series resonant circuits have low inductive voltages.

T F **3.** Excessive voltage in a series resonant circuit can damage components.

T F **4.** At resonant frequency, a series resonant circuit is neither capacitive nor inductive.

T F **5.** In a series resonant circuit, minimum power occurs at resonance.

T F **6.** A series resonant circuit with a high quality factor has high resistance and low current.

T F **7.** A parallel resonant circuit has an angle theta equal to 270°.

T F **8.** Voltage is not measured in a parallel resonant circuit because it is typically known.

T F **9.** Total circuit impedance in a parallel resonant circuit can be calculated when source voltage and total current are known.

T F **10.** The characteristics of a frequency filter make it possible to select broadcast from one radio station while excluding others.

T F **11.** Frequency filters are used to create harmonics.

T F **12.** A low-pass frequency filter blocks all frequencies above a selected frequency.

T F **13.** Band-pass filters are usually used in communications applications to tune frequency bands.

T F **14.** Band-reject frequency filters increase signals between selected frequencies.

Multiple Choice

_______ 1. Resonance is the condition where the inductive ___ equals capacitive ___ at a given frequency.
   A. reactance; reactance
   B. voltage; current
   C. current; current
   D. resistance; reactance

_______ 2. Impedance in a series resonant circuit is equal to the vector sums of reactance and ___.
   A. current
   B. frequency
   C. voltage
   D. resistance

_______ 3. ___ is the range of frequencies that the circuit passes without a significant reduction in the signal magnitude.
   A. Bandwidth
   B. Harmonics
   C. Quality factor
   D. Power factor

_______ 4. The ___ of a parallel resonant circuit is the ratio of the tank current to the total current.
   A. power factor
   B. quality factor
   C. resistance
   D. none of the above

_______ 5. Frequency filters have all the following components except ___.
   A. resistors
   B. capacitors
   C. shunts
   D. inductors

_______ 6. A ___ is a frequency that is an integer multiple of the fundamental frequency.
   A. harmonic
   B. filter
   C. half-power point
   D. none of the above

_______________ **7.** ___ is a reduction in the strength of a frequency signal.
A. Bandwidth
B. Harmonics
C. Attenuation
D. Cutoff frequency

_______________ **8.** A ___ frequency is the half-power frequency point.
A. cutoff
B. low
C. high
D. band-pass

_______________ **9.** A ___ network is a system of band-pass filters used to route the appropriate signal to the appropriate loudspeaker.
A. frequency
B. crossover
C. harmonic
D. bandwidth

_______________ **10.** A frequency filter ___ is a complex design of frequency filters arranged to improve filter performance.
A. system
B. application
C. diagram
D. network

Completion

_______________ **1.** ___ is used as the reference point in a series resonant circuit.

_______________ **2.** In a series resonant circuit, ___ across the inductor and capacitor can exceed source voltage.

_______________ **3.** The frequency where circuit power is half of the resonant power is referred to as the ___.

_______________ **4.** ___ is used as the reference point in a parallel resonant circuit.

_______________ **5.** Parallel resonant circuits are sometimes referred to as ___ circuits because of their ability to store energy.

_______________ **6.** A(n) ___ sine wave is a sine wave that has been reduced due to a resistance consuming power.

__________ **7.** Frequency filter types are low-pass, high-pass, band-pass, and ___.

__________ **8.** A(n) ___ frequency filter is one that allows all frequencies above a selected frequency to be applied to a load.

__________ **9.** Reject frequencies are frequencies rejected because of ___ or size.

__________ **10.** A(n) ___ frequency filter passes signals between selected frequencies.

Identification—Pi and T Networks

__________ **1.** Band-reject

__________ **2.** Band-pass

__________ **3.** Low-pass

__________ **4.** High-pass

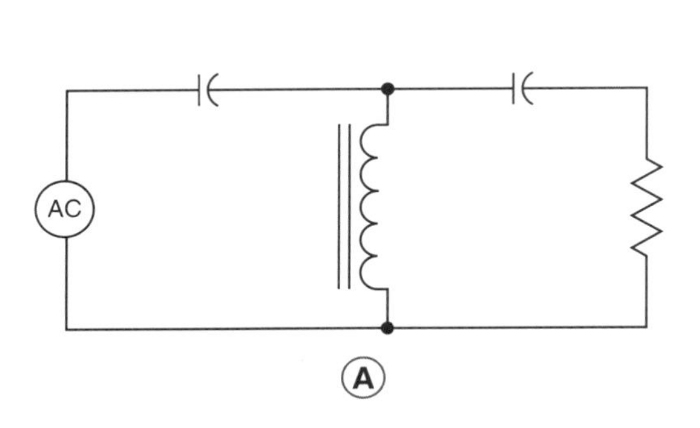

Ⓐ

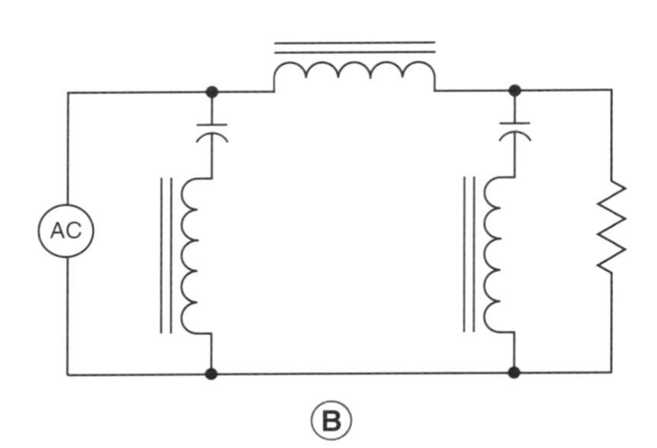

Ⓑ

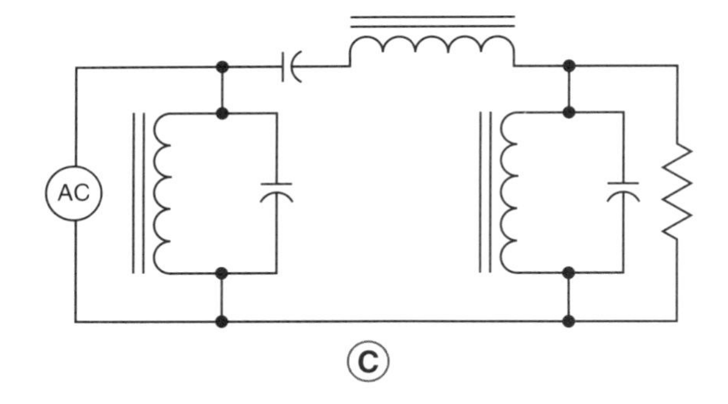

Ⓒ

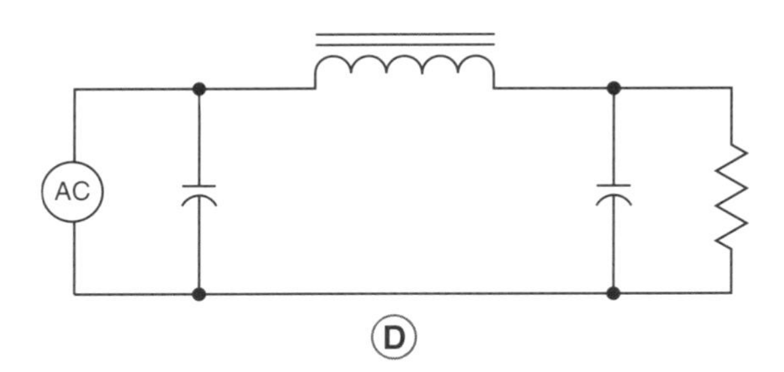

Ⓓ

Series Resonant Circuit Calculations

Calculate the voltage drop across each component in a series resonant circuit with a total current of 0.8 A, an inductive and capacitive reactance of 800 Ω, and a resistance of 150 Ω.

________________ **1.** V_L = ___ V

________________ **2.** V_C = ___ V

________________ **3.** V_R = ___ V

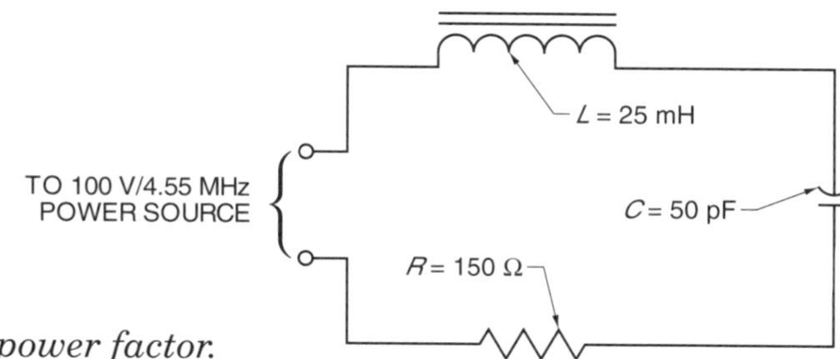

Calculate the apparent power, true power, and power factor.

________________ **4.** P_A = ___ VA

________________ **5.** P_T = ___ W

________________ **6.** PF = ___%

________________ **7.** Is the circuit at resonance?

Calculate the voltage drop across each component in a series resonant circuit with a total current of 0.4 A, an inductive and capacitive reactance of 720 Ω, and a resistance of 300 Ω.

________________ **8.** V_L = ___ V

________________ **9.** V_C = ___ V

________________ **10.** V_R = ___ V

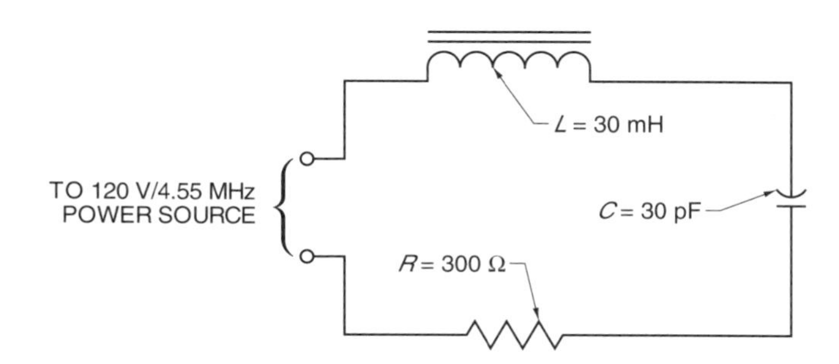

Calculate the apparent power, true power, and power factor.

______________ **11.** P_A = ___ VA

______________ **12.** P_T = ___ W

______________ **13.** PF = ___%

______________ **14.** Is the circuit at resonance?

Parallel Resonant Circuit Calculations

Calculate the total line current and angle theta in a parallel resonant circuit with an inductive branch current of 0.991 A at –85° and a capacitive branch current of 1.1 A at 90°.

______________ **1.** I_{VL} = ___ A

______________ **2.** I_{VC} = ___ A

______________ **3.** I_{HL} = ___ A

______________ **4.** I_{HC} = ___ A

______________ **5.** I_T = ___ A

______________ **6.** θ = ___°

______________ **7.** Z_T = ___ Ω at ___°

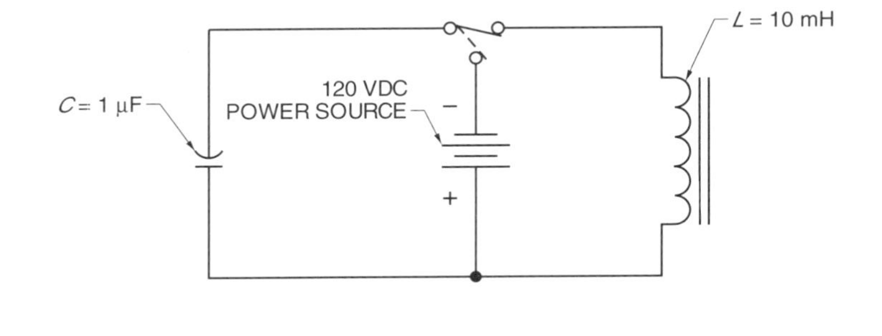

Calculate the apparent power, power factor, and true power.

______________ **8.** P_A = ___ VA

______________ **9.** PF = ___%

______________ **10.** P_T = ___ W

Calculate the total line current and angle theta in a parallel resonant circuit with an inductive branch current of 0.9. A at 90° and a capacitive branch current of 0.996 A at –86°.

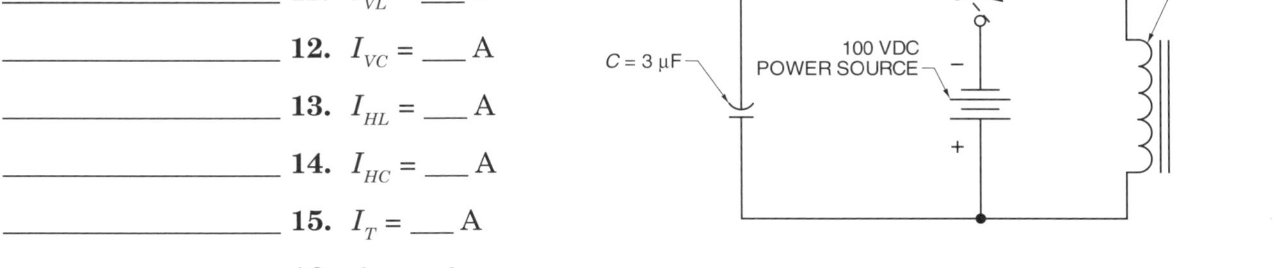

_______________ **11.** I_{VL} = ___ A

_______________ **12.** I_{VC} = ___ A

_______________ **13.** I_{HL} = ___ A

_______________ **14.** I_{HC} = ___ A

_______________ **15.** I_T = ___ A

_______________ **16.** θ = ___°

_______________ **17.** Z_T = ___ Ω at ___°

Calculate the apparent power, power factor, and true power.

_______________ **18.** P_A = ___ VA

_______________ **19.** *PF* = ___%

_______________ **20.** P_T = ___ W

Low-Pass Filters

A low-pass frequency filter allows all frequencies below a selected frequency to be applied to a load. All frequencies above that point are blocked. Low-pass filters may also be used to remove a high frequency joined to a lower frequency.

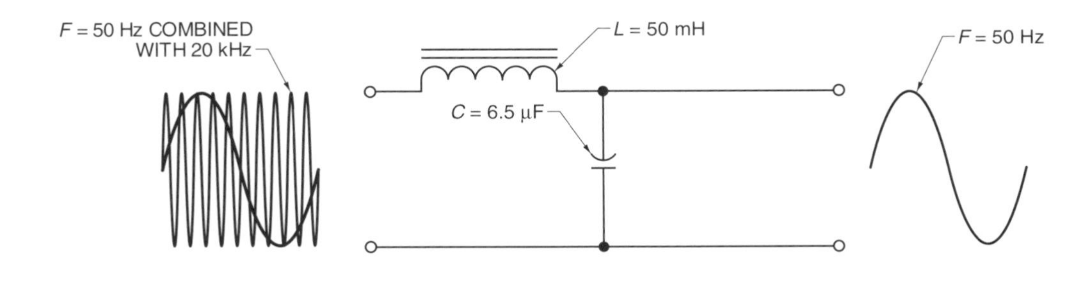

Calculate the inductive reactance at each frequency.

_______________ **1.** X_L (50 Hz) = ___ Ω

_______________ **2.** X_L (20 kHz) = ___ kΩ

Calculate the capacitive reactance at each frequency.

_______________ **3.** X_C (50 Hz) = ___ kΩ

_______________ **4.** X_C (20 kHz) = ___ Ω

High-Pass Filters

High-pass filters are often used to allow frequencies above a selected frequency to be applied to a load. All frequencies below that point are blocked. A reactive voltage divider can be used with a high-pass filter. Using a reactive voltage divider changes the orientation of the components within the circuit. In a high-pass filter with a voltage divider, the capacitor is connected in series, and the resistor is in parallel with the voltage.

Rearrange the capacitive reactance formula to calculate the cutoff frequency.

_______________ **1.** f = ___ Hz

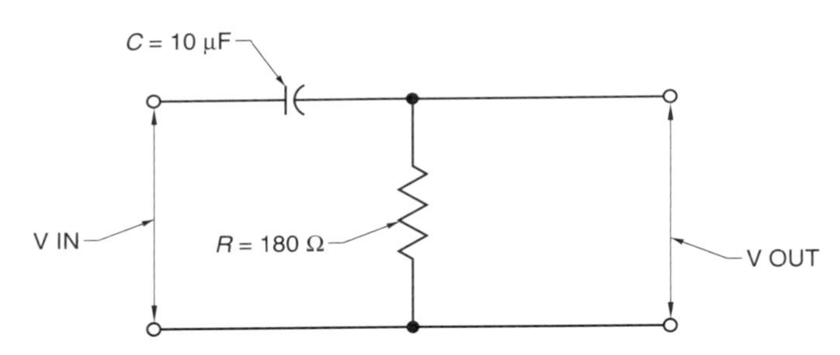

Now consider a high-pass filter with an inductor in parallel with the voltage.

Rearrange the capacitive reactance formula to calculate the cutoff frequency.

_______________ **2.** f = ___ Hz

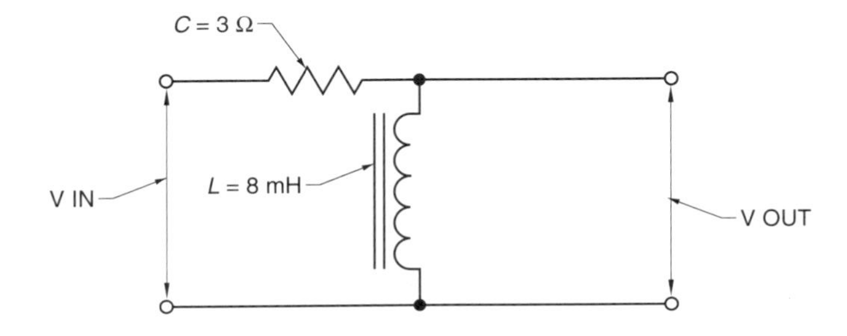

Three-Phase AC 19

Name ______________________________ Date ______________

True-False

T F 1. Three-phase power is often preferred to single-phase power because less copper is used to deliver a given amount of power.

T F 2. Single-phase power creates smoother DC output with less filtering than three-phase power.

T F 3. Three-phase motors require more components to operate and often require more maintenance than 1ϕ motors.

T F 4. A rotor is the rotating part of an AC motor.

T F 5. An AC generator cannot produce output voltage before the rotor has been excited.

T F 6. General voltage is a voltage produced around a closed path or circuit by a change in the magnetic lines of force linking the path.

T F 7. Clamp-on ammeters measure current by measuring the strength of the voltage around a single conductor.

T F 8. Appliances that produce heat or provide incandescent light are examples of resistive loads.

T F 9. The line currents of a delta-connected generator have equal values and are located 120° apart.

T F 10. Generated voltages in a delta-connected generator have different values from the load voltages.

T F 11. In a wye connection, five leads are connected together, while the other five leads are brought out for connecting to the load.

T F 12. A wye connection is sometimes referred to as a star connection.

T F 13. A wye-connected generator has line voltages that are equal in value.

T F 14. Each branch of a delta connection has the same value of impedance.

Multiple Choice

________ **1.** Three-phase power systems use three individual ___ to transmit power.
A. inductors
B. conductors
C. resistors
D. transformers

________ **2.** ___ are used to produce three-phase power.
A. Generators
B. Carburetors
C. Coils
D. none of the above

________ **3.** A(n) ___ is a movable coil of wire in a generator that rotates relative to the magnetic field.
A. engine
B. motor
C. winding
D. armature

________ **4.** A(n) ___ resistor is a resistor that is used to suppress the high inductive kick of an exciter current.
A. swamping
B. paper
C. exciting
D. ceramic

________ **5.** A(n) ___ is a test instrument that is used to measure true power in a circuit.
A. ammeter
B. wattmeter
C. power quality meter
D. voltage meter

________ **6.** Power factor is lagging for an inductive load, leading for a capacitive load, and ___ for a resistive load.
A. lagging
B. leading
C. in phase
D. none of the above

_______________ **7.** Power factor correction is the process of correcting power factor to at least ___%.
A. 70
B. 80
C. 85
D. 90

_______________ **8.** ___ is the frequent starting and stopping of a motor for short periods.
A. Plugging
B. Driving
C. Jogging
D. Loading

_______________ **9.** A power-correcting capacitor does all of the following except ___.
A. reduce line losses
B. increase current levels
C. increase system capacity
D. improve voltage levels

_______________ **10.** A ___ connection is a connection that has each coil connected end-to-end to form a closed loop.
A. delta-to-delta
B. wye-to-delta
C. wye
D. delta

_______________ **11.** A(n) ___ circuit is either capacitive or inductive and has a power factor of less than 1.
A. capacitive
B. reactive
C. inductive
D. none of the above

_______________ **12.** A flywheel ___ is used in an alternator to suppress the inductive voltage that occurs when the exciter switch is opened.
A. fuse
B. switch
C. diode
D. motor

_______________ **13.** Which of the following is not an example of an inductive load?
A. motor under full load
B. transformer
C. induction furnace
D. motor without load

Completion

______ **1.** In a three-phase power system, the voltage of each conductor is ___ out of phase with that of the other conductors.

______ **2.** A(n) ___ is a device that converts AC voltage to DC voltage by allowing the AC voltage and current to flow in only one direction.

______ **3.** Metallic rings that are connected to the ends of the armature and are used to connect the induced voltage to the brushes are known as ___ rings.

______ **4.** A(n) ___ is the number of completely isolated circuits that a relay can switch.

______ **5.** Power ___ is the ratio of true power used in an AC circuit to apparent power delivered to the circuit.

______ **6.** "Alternator" is another name for an AC ___.

______ **7.** ___ is a method of motor braking in which the motor connections are reversed so that the motor develops a countertorque that acts as a braking force.

______ **8.** A(n) ___ connection is a connection that has one end of each coil connected together and the other end of each coil left open for external connections.

______ **9.** The ___ of a coil is how tightly the windings of a coil are wound.

______ **10.** A(n) ___ is the sliding contact that rides against the slip rings and is used to connect the armature to the external circuit.

Power Factor Analysis

What is the power factor analysis for a 650 V, 40 A circuit that has a 30 HP motor and needs to be operating with a power factor of 0.93 (93%)?

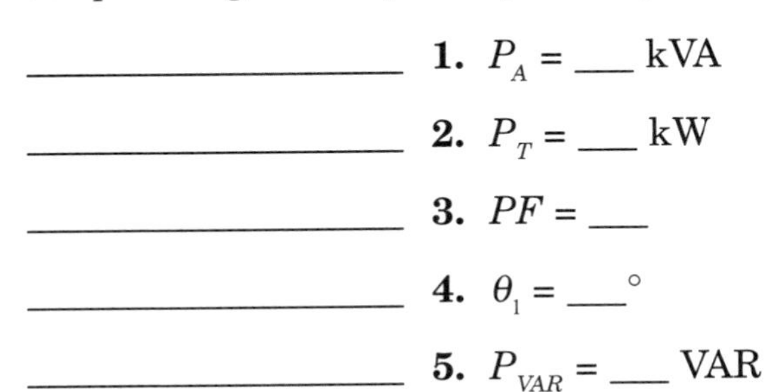

______ **1.** P_A = ___ kVA

______ **2.** P_T = ___ kW

______ **3.** PF = ___

______ **4.** θ_1 = ___°

______ **5.** P_{VAR} = ___ VAR

__________________ **6.** $\theta_2 =$ ___°

__________________ **7.** $CkVAR =$ ___ CkVAR

__________________ **8.** $P_A =$ ___ kVA

__________________ **9.** $I_{LINE} =$ ___ A

What is the power factor analysis for a 500 V, 30 A circuit that has a 20 HP motor and needs to be operating with a power factor of 0.96 (96%)?

__________________ **10.** $P_A =$ ___ kVA

__________________ **11.** $P_T =$ ___ kW

__________________ **12.** $PF =$ ___

__________________ **13.** $\theta_1 =$ ___°

__________________ **14.** $P_{VAR} =$ ___ VAR

__________________ **15.** $\theta_2 =$ ___°

__________________ **16.** $CkVAR =$ ___ CkVAR

__________________ **17.** $P_A =$ ___ kVA

__________________ **18.** $I_{LINE} =$ ___ A

Power Factor

Consider a 3ϕ motor used to pump latex paint from a storage tank into paint pails on an assembly line. The 20 HP motor has a power factor of 0.88.

Determine the size of capacitor needed to correct the power factor to 0.95 by calculating CkVAR.

__________________ **1.** $CkVAR =$ ___ CkVAR

What is the apparent power after the system is corrected?

__________________ **2.** $P_A =$ ___ kVA

Power Consumption

Consider an industrial heater used to heat plastic in an injection-molding machine. The heater uses a 3ϕ motor and 3 resistors. Each resistor has a value of 10 Ω. The voltage applied is 240 V.

Use Ohm's law to calculate the current within the system.

____________________ **1.** $I =$ ___ A

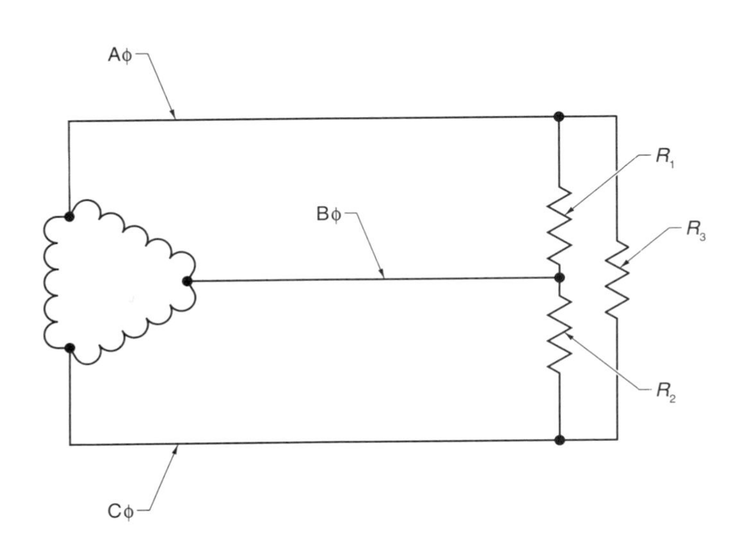

Assume that the heater has 18 HP. Compare the values of apparent power and true power of the heater.

____________________ **2.** $P_A =$ ___ kVA

____________________ **3.** $P_T =$ ___ kW

What is the power factor?

____________________ **4.** $PF =$ ___

Transformers

Name ______________________________ Date ______________

True-False

T F **1.** A wall transformer has prongs for connection to an electrical receptacle.

T F **2.** Most transformers have only one winding.

T F **3.** Increasing voltage and reducing current in a system can reduce the cost of power lines.

T F **4.** Voltage and current ratings of a transformer must be known to calculate power ratings.

T F **5.** Overheating of a transformer often occurs when the transformer is operating above its secondary voltage.

T F **6.** When a transformer is connected to an inductive-resistive-capacitive load, the circuit is purely resistive at resonance.

T F **7.** The turns ratio is the ratio of the number of turns in the secondary winding to the number of turns in the primary winding.

T F **8.** In a step-up transformer, current is stepped down.

T F **9.** Impedance matching adjusts the load-circuit impedance to produce the desired energy transfer from the power source to the load.

T F **10.** Current flow through the windings of a transformer can generate heat loss.

T F **11.** Induced currents in an iron core are desirable because they produce heat and use power.

T F **12.** Transformer efficiency is the ratio of the output current to the input current.

T F **13.** The frequency at which a transformer operates determines its classification.

T F **14.** A multiple-winding transformer has more than four windings on its core.

T F **15.** One advantage of using an autotransformer rather than other types of power transformers is that current flow through the windings is reduced.

T F **16.** Buck-boost transformers can lower or raise line voltage.

T F **17.** Voltage across any winding of a transformer has polarity.

T F **18.** Because core-type transformers use more iron in the core than other transformers, they are usually more expensive.

T F **19.** A potential transformer is a type of instrument transformer.

T F **20.** A balun is a type of buck-boost transformer.

Multiple Choice

______________ **1.** A transformer is an electrical device that uses ___ to change voltage from one level to another or to isolate one voltage from another.
- A. chemistry
- B. resistance
- C. electromagnetism
- D. static electricity

______________ **2.** A(n) ___ transformer has equal primary and secondary voltages and is used to electrically separate the source from the load.
- A. isolation
- B. wall
- C. step-up
- D. step-down

______________ **3.** Mutual inductance, the measure of efficiency with which power is transferred between windings, is also known as ___.
- A. exciting current
- B. coefficient of coupling
- C. turns ratio
- D. none of the above

______________ **4.** A transformer that is mainly inductive has which of the following characteristics?
- A. little current drawn from the circuit
- B. primary voltage equal to source voltage
- C. secondary voltage connected to large resistive load
- D. all of the above

_______________ **5.** Impedance matching is also known as ___ matching.
A. load
B. energy
C. transformer
D. power

_______________ **6.** ___ loss is power loss in a transformer caused by the resistance of the copper conductors used to make the windings.
A. Energy
B. Flux
C. Current
D. Resistive

_______________ **7.** ___ loss is power loss in a transformer or motor due to currents induced in the metal field structure from the changing magnetic field.
A. Hysteresis
B. Eddy current
C. Flux
D. all of the above

_______________ **8.** Which of the following is classified as a power transformer?
A. autotransformer
B. buck-boost
C. tapped-winding
D. all of the above

_______________ **9.** An autotransformer is a transformer in which the primary and secondary circuits have portions of their two ___ in common.
A. windings
B. cores
C. voltages
D. all of the above

_______________ **10.** A(n) ___ transformer is a transformer that has windings placed around each leg of the core material.
A. power
B. isolation
C. core-type
D. buck-boost

_______________ **11.** A(n) ___ transformer is a precision two-winding transformer that is used to step down high voltage to allow safe voltage measurement.
A. instrument
B. potential
C. audio
D. none of the above

_______________ **12.** Which connection is not possible for a three-phase transformer?
A. delta-delta
B. wye-wye
C. wye-delta
D. none of the above

Completion

_______________ **1.** In a transformer schematic symbol, an "H" indicates the primary winding and a(n) "___" indicates the secondary winding.

_______________ **2.** If the turns ratio of primary to secondary windings is 2:1, the transformer is a(n) ___ transformer.

_______________ **3.** ___ is the number of flux lines linking the primary and secondary transformer windings.

_______________ **4.** Current ratings in a transformer are usually specified for the ___ windings only.

_______________ **5.** The ___ load of a transformer is the primary impedance of the transformer that is directly proportional to the secondary load.

_______________ **6.** ___ is the voltage dropped across each turn of a winding or the voltage induced into each turn of the secondary winding.

_______________ **7.** The property where the magnetic induction of a winding lags the magnetic field that is charging the winding is known as ___.

_______________ **8.** Flux loss occurs when some of the ___ from the primary winding do not travel through the core to the secondary winding.

_______________ **9.** A(n) ___ transformer is used to raise or lower voltage as required to serve transmission circuits.

_______________ **10.** An isolation transformer has complete ___ separation between the primary and secondary windings.

Calculating Reflected Impedance of a Transformer

Calculate the reflected impedance of a transformer that has a turns ratio of 4:1, a secondary inductance of 250 Ω, and a secondary resistance of 300 Ω.

_______________ **1.** X_{RP} = ___ Ω

_______________ **2.** R_P = ___ Ω

_______________ **3.** Z_P = ___ Ω

Calculate the reflected impedance of a transformer that has a turns ratio of 2:1, a secondary inductance of 400 Ω, and a secondary resistance of 600 Ω.

_______________ **4.** X_{RP} = ___ Ω

_______________ **5.** R_P = ___ Ω

_______________ **6.** Z_P = ___ Ω

Calculate the reflected impedance of a transformer that has a turns ratio of 3:1, a secondary inductance of 550 Ω, and a secondary resistance of 750 Ω.

_______________ **7.** X_{RP} = ___ Ω

_______________ **8.** R_P = ___ Ω

_______________ **9.** Z_P = ___ Ω

Transformer Calculations

Answer questions 1–8 using the circuit below.

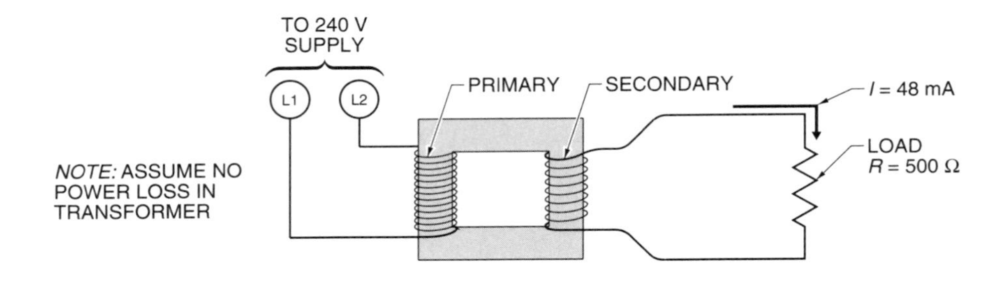

______ **1.** Secondary voltage = ___ V

______ **2.** Transformer voltage ratio = ___ : ___

______ **3.** Transformer current ratio = ___ : ___

______ **4.** Primary current = ___ mA

______ **5.** Primary power = ___ mW

______ **6.** Secondary power = ___ mW

______ **7.** Transformer power ratio = ___ : ___

______ **8.** Transformer turns ratio = ___ : ___

Answer questions 9–16 using the circuit below.

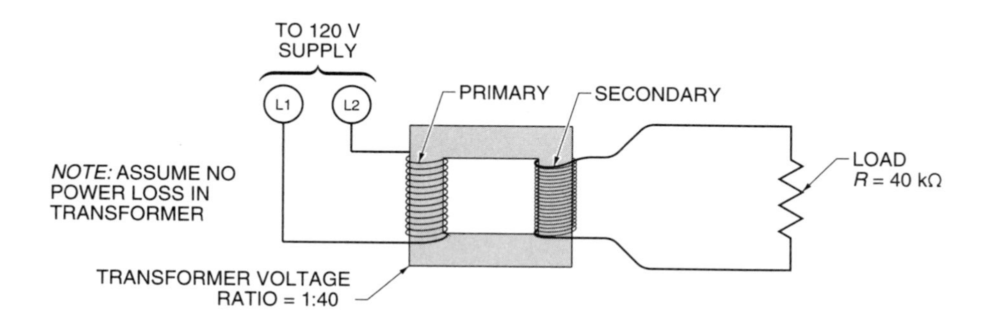

____________________ **9.** Secondary voltage = ___ V

____________________ **10.** Secondary current = ___ mA

____________________ **11.** Primary current = ___ A

____________________ **12.** Primary power = ___ W

____________________ **13.** Secondary power = ___ W

____________________ **14.** Transformer current ratio = ___ : ___

____________________ **15.** Transformer power ratio = ___ : ___

____________________ **16.** Transformer turns ratio = ___ : ___

Transformer Capacity and Current Draw

To calculate the minimum kVA capacity needed for a 3ϕ transformer when voltage and current are known, the following formula is used:

$$kVA_{CAP} = V \times \frac{I}{1000}$$

where

kVA_{CAP} = transformer capacity (in kVA)

V = voltage (in V)

1.732 = constant (√3)

I = current (in A)

1000 = constant

Use the capacity formula to answer the following problems:

________________ **1.** A 3ϕ transformer, 240 V, with loads of 25 A, 30 A, and 8 A, has a kVA capacity of ___ kVA.

________________ **2.** A 3ϕ transformer, 240 V, with loads of 20 A, 40 A, and 10 A, has a kVA capacity of ___ kVA.

________________ **3.** A 3ϕ transformer, 240 V, with loads of 45 A, 22 A, and 13 A, has a kVA capacity of ___ kVA.

Determining Three-Phase Transformer Current Draw

To calculate the current draw of a 3ϕ transformer when voltage and kVA capacity are known, use the following formula:

$$I = kVA_{CAP} \times \frac{1000}{V \times 1.732}$$

where

I = current (in A)

kVA_{CAP} = transformer capacity (in kVA)

1000 = constant

V = voltage (in V)

1.732 = (√3)

Use the capacity formula to answer the following problems:

________________ **1.** A 30 kVA, 480 V, 3ϕ transformer, fully loaded, draws ___ A.

________________ **2.** A 22 kVA, 240 V, 3ϕ transformer, fully loaded, draws ___ A.

________________ **3.** A 45 kVA, 480 V, 3ϕ transformer, loaded at 75%, draws ___ A.

AC Motors 21

Name ______________________ Date ____________

True-False

T	F	**1.** Less maintenance is required for a 1ϕ motor than for a 3ϕ motor.
T	F	**2.** Changing any two connections of a 3ϕ motor will cause the rotation of the magnetic field to reverse.
T	F	**3.** As slip increases, torque decreases.
T	F	**4.** A cage rotor is sometimes referred to as a "squirrel-cage rotor".
T	F	**5.** A variable resistance is connected in series with rotor windings of a wound rotor to limit inrush current.
T	F	**6.** A synchronous motor is a 1ϕ motor that runs at synchronous speed.
T	F	**7.** Synchronous motors are self-starting.
T	F	**8.** A 1ϕ induction motor is comparable to a transformer.
T	F	**9.** A split-phase motor is a single-phase AC motor with a running winding and a starting winding.
T	F	**10.** If the centrifugal switch of a split-phase motor opens, the windings will overheat and destroy the motor.
T	F	**11.** Split-phase motors can operate with two separate voltages.
T	F	**12.** Centrifugal switches are required in capacitor-run motors.
T	F	**13.** In a hard neutral position, a rotating magnetic field is not present.
T	F	**14.** Repulsion motors should always be started from the soft neutral position.

Multiple Choice

______________ **1.** Which of the following characteristics describes a 3ϕ induction motor?
A. can vary motor speed
B. uses one of two different types of rotors
C. can switch motor rotation
D. all of the above

______________ **2.** ___ is the difference between the synchronous speed and the actual speed of a motor.
A. Power
B. Torque
C. Slip
D. Flow

______________ **3.** The two types of rotors used on AC motors are the cage rotor and the ___ rotor.
A. squirrel-cage
B. wound
C. brush
D. none of the above

______________ **4.** Which of the following is not an advantage of a wound-rotor induction motor over a cage motor?
A. reduced heating of rotor during starting
B. lower initial cost
C. adjustable speed control
D. smooth acceleration under heavy loads

______________ **5.** A(n) ___ switch opens to disconnect the starting winding when the rotor reaches a preset speed and reconnects when the speed falls below a set value.
A. centrifugal
B. open
C. starter
D. inductive

______________ **6.** In the construction of a capacitor motor, a capacitor is connected in ___ with the starting winding.
A. series
B. parallel
C. series/parallel
D. none of the above

________________ **7.** A ___ pole is a single pole, normally made of a single turn of heavy-gauge copper wire, used for starting a single-phase motor.
A. north
B. south
C. shaded
D. salient

________________ **8.** A(n) ___ motor is a motor with the rotor connected to the power supply through brushes that ride on the commutator.
A. AC
B. DC
C. stator
D. repulsion

________________ **9.** A ___ neutral position is a position where the brushes are aligned 90° from the stator poles.
A. soft
B. hard
C. single
D. none of the above

________________ **10.** A compensation winding is an inductor used to reduce ___ reactance effects.
A. capacitive
B. armature
C. inductive
D. all of the above

Completion

________________ **1.** The armature of a motor will only turn when it is in the rotating magnetic field caused by the ___.

________________ **2.** ___ is the theoretical speed of a motor based on line frequency and the number of poles of the motor.

________________ **3.** Wound-rotor windings are typically ___ connected.

________________ **4.** Limiting inrush current gives a motor greater startup ___ at a lower startup current.

________________ **5.** Magnetic polarity is based on the ___ of current flow.

________________ **6.** The windings of a split-phase motor are connected in ___.

_______________ **7.** Capacitor motors range in size from ⅛ HP to ___ HP.

_______________ **8.** A capacitor-___ motor has the most running torque of all the capacitor-style motors.

_______________ **9.** A salient pole is a pole that consists of a separate radial projection having its own iron pole and ___.

_______________ **10.** A(n) ___ motor is an AC series motor with brushes and a wound armature.

Induction Motors

A frequency of 50 Hz is applied to a motor being used to dry textiles in preparation for shipment and storage. The motor is a 3ϕ induction motor with 4 poles.

What is the synchronous speed?

_______________ **1.** Ω_s = ___ rpm

Slip is the difference between the synchronous speed and the actual speed of a motor. The amount of slip affects the true speed of the motor. In an ideal motor, there would be no slip and synchronous speed would equal the actual speed. In real world applications, all motors have slip. As slip increases, the true speed of the motor decreases. The slip for this motor is 4%.

What is the true speed of the motor?

_______________ **2.** Ω_t = ___ rpm